TRACK TECHNOLOGY

TRACK TECHNOLOGY

Proceedings of a conference organized by the
Institution of Civil Engineers and held at the
University of Nottingham, 11 – 13 July 1984

THOMAS TELFORD LTD, LONDON

Published for the Institution of Civil Engineers by Thomas Telford Ltd, PO Box 101, 26 – 34 Old Street, London EC1P 1JH.

First published 1985.

Organizing Committee: R. H. Busby, R. J. Collins, D. S. Currie, D. G. Jobling, R. W. Sparrow.

British Library Cataloguing in Publication Data
Track technology: proceedings of a conference
    organized by the Institution of Civil Engineers
    and held at the University of Nottingham,
    11–13 July 1984.
    1. Railroads — Track   2. Railroad engineering
    I. Institution of Civil Engineers
    625.1'4   TF240

ISBN: 0 7277 0228 9

Printed in Great Britain by Billing & Sons Ltd, Worcester.

Contents

Opening address

J. A. GAFFNEY, President, The Institution of Civil Engineers

Ladies and gentlemen, firstly I would like to thank our Chairman, Mr Currie, for his kind introduction and to welcome you most heartily to the conference.

Looking around me, I am interested and pleased to note the number of young faces in the audience and this puts me in mind of the early objectives of the Institution of Civil Engineers. The Institution's role as a gathering of members became a learned society, the development of which has included conferences such as this.

The Institution was formed by a group of young men in the early 19th century who wished to gain from the experience of others and not just from their masters. The all-embracing title 'Civil Engineer' was used to differentiate from military engineering. From the earliest times public works which, by their nature, required a regulated labour force were undertaken by kings and armies and the captives they enslaved or serfs under direction. The early days of the Institution embraced the era of canals, aqueducts, irrigation and toll roads, but the railway age was upon us.

There have been numerous distinguished contributions to the Institution's activities and proceedings from railway engineers and, particularly, many eminent presidents. These contributions have now served and benefitted society's needs and development for over a century.

Although I come from 'the other side' in a manner of speaking, it gives me great personal pleasure to be here at this conference because of Nottingham connections. Transportation is now considered as a whole discipline which includes all forms, as well as highways (which are my background, of course, during the motorway era) - not, I should add, that more traffic on the roads is wanted. The Institution has its transportation engineering group which is representative of all forms of transport.

The conference in Hong Kong on transportation, jointly sponsored by the Institution of Civil Engineers a couple of years ago, demonstrated that the Institution has a large international membership. The overseas delegates (from Japan, Hong Kong, Canada, Australia etc.) are very welcome to our proceedings and it is a great pleasure to meet them.

So we now come on to the objectives of this conference.

The first objective is to share experience. Sharing experience in technical development and knowledge will naturally be of great benefit to all.

Secondly is the balance in content. It is important to balance the practical contributions from research and analysis with experience, which contribute to discussion and debate.

Finally is the importance of the topic. Rail travel (both freight and passenger) is of international importance today. Society is becoming more and more urban which is creating greater servicing and safety requirements, and the safe operation of the railways is well known and admired. There are many factors involved but particularly proper track provision and good maintenance.

In conclusion, then, it is a great pleasure for me to be here as President of the Institution in the centenary year of the Permanent Way Institute, which started in Nottingham, and to have the support of Mr Currie (a past chairman of the Transportation Board). Perhaps in earlier times the Institution did not succeed fully in catering for specializations but it is now (and has been for some while) determined to include all relevant topics in its learned society programmes.

I now formally open this conference on Track Technology for the Next Decade.

1 Railway track in a competitive commercial business

I. M. CAMPBELL, CVO, BSc, FEng, FICE, Member, British Railways Board, London, UK

During the three days of this conference the technology of track construction, renewal and maintenance from many parts of the world, will be discussed for many different types of rail traffic and operation and under many different financial and regulatory conditions.

One thing is common to all of us. Track is expensive and is a major factor in the economics of competition of a successful railway business. In the U.K. for every £100 spent in providing the end product £25 is spent on the infrastructure - track, structures and signalling. Of this, £13 is for track alone. In absolute terms this total track cost in 1984 will be about £330m.

Two conclusions can be drawn from these simple figures. First, if the track is used only sparsely, the cost per train unit and hence per passenger or per ton of freight is likely to lead to diminishing competitiveness in the transport market. The second conclusion, which concerns us at this conference, is that, in dealing with such large sums of money, it is necessary to apply the highest standard of professional and technical skill to decrease unit costs and apply the best of technology to optimise productivity.

Perhaps these truths are self evident but they may be illustrated by some over simplified figures as set out in Appendix A.

These are simple figures and no doubt their relevance and accuracy can be debated. Nevertheless the difference between road and rail infrastructure costs is highly relevant. The road operator, apart from a minute annual licence fee, pays only for what he uses, through fuel tax. The rail infrastructure cost is, within small limits, constant and the cost per passenger or per ton of freight varies according to the volume of traffic.

The figures for road and rail are not, of course, strictly comparable. They assume the doubtful hypothesis that all fuel tax should be dedicated to the road infrastructure. They take no account of policing and regulation, or of tax benefits to the commercial private car. They assume equal standards of safety and the cost of achieving such standards. They take no account of the less finite

effects on the environment. They take no account of terminal
provision. If these were taken into the equation it might
be concluded that the terms of competition are by no means
equal. This is not, however, a conference to debate the
economics of road versus rail.

Nevertheless, the figures are an indication of the
large cost of track, structures and signalling together with
train operating costs which have to be met by the commercial
sellers of rail transport in a competitive market. Cost
effectiveness of the provision of track is a vital
ingredient.

To this end, it is important to define the infrastructure requirements for speed and tonnage which will meet
the requirements of the different sectors of the business
for each route at realistic levels for the foreseeable
future.

This conference will be concerned with discussing the
best of today's practice and the developments which the
future will bring. I am not competent to enter into that
discussion but it is useful to look back some years and
consider the changes which have taken place. I choose a
period of 30 years, first because it encompasses the period
of my own experience and secondly because it covers a period
of many changes, changes in materials, in the use of
machines and men, in the conditions of employment of manpower
and in the demands made upon track by much higher speeds and
higher axle loads.

The changes described apply to British Railways
specifically, but apply in general terms to most mixed
traffic systems throughout the world. In Britain there has
been no significant construction of new trunk lines and all
these changes have been carried out on a network largely
determined in the nineteenth century, which has been adapted
to meet modern requirements by improving alignment and
simplifying junction and station layouts. Most of the basic
infrastructure, such as embankments, cuttings, bridges and
tunnels remains unchanged.

British Railways in the early 1950's

A network of 19,150 route miles (30,800kms.), 35,500
track miles (57,150kms.) which was mostly intensively used,
but including many miles of lightly used rural lines.

All trains were hauled by steam locomotives which were
generally the heaviest vehicle with axle loads up to $22\frac{1}{2}$ tons.
Passenger speeds were no higher than 90m.p.h. (140k.p.h.),
most routes with lower speeds. Freight trains operated at
lower speeds most at no more than 50m.p.h. (80k.p.h.). There
was a large fleet of wagons, most of two axles limited to
20 tons axle loads.

The track layout on most trunk routes was complex
particularly in main junction areas with a multitude of turn-
outs, diamonds and slips, many located on curves.

Track material consisted of 95lb./yd. (47kg./m.) Bull

head rail with the complexity of keys, chairs and screws
seated upon treated softwood sleepers of variable quality.
The ballast was of variable quality, from the best of igneous
rock to the very variable quality of limestone, slag or ash.
Whilst ballast usually appeared to be of good quality on the
surface, the sub-ballast was usually almost impervious,
contaminated by locomotive ash and the natural intrusion of
the sub-base.   There were no facilities, apart from reconstruction, to clean ballast to a practical depth below sleeper
bottom for good drainage.   The laborious task of manual
screening of ballast was carried out only intermittently.

Relaying and maintenance were carried out almost
entirely by hand.   There were few machines available, apart
from cranes, to assist manpower.   The extent of mechanisation
was the installation of 60ft. (18.3m.) panels of prefabricated track by crane.

The track worker was, by tradition, prepared to accept
these heavy tasks, carried out in all weathers, night and day
for 48 hours a week plus voluntary weekends.   The skill of
the worker, despite training, was very variable, a few only
being able to carry out high quality skilled measured
shovel packing.

All track was jointed, requiring annual lubrication of
joints and visual inspection for rail failure in the weakest
part of the track structure.

In retrospect, these may appear to be very primitive
conditions under which to operate one of the most intensive
rail systems in the world.   And yet it worked well despite
the lack of investment of any magnitude for the previous
15 years, but at a high cost.

The 30 years between have seen major changes to meet
new traffic demands, new conditions of employment, new methods
of work measurement in which British Railways were pioneers
and new opportunities in the use of purpose designed plant and
machinery to assist, or in some cases, to replace the
limitations of human muscle.

British Railways in 1984

A substantial reduction of network size has taken
place down to 11,000 route miles (17,700kms.), 20,480 track
miles (32,960kms.).   Much of the closure has been of the
rural and freight lines where standards of materials and
manning levels are lower than for the main core of the
network which remains.

All trains are now hauled by electric or diesel
locomotives.   Passenger speeds on 5% of the total network
are as high as 125m.p.h. (200k.p.h.) and on 15% of the network
are as high as 100m.p.h. (160k.p.h.).   The slower freight
trains have been largely eliminated on the trunk routes.
Freight moves in train loads rather than wagon loads
travelling at speeds up to 75m.p.h. (120k.p.h.) mostly in
larger wagons with a much improved ratio of payload to
unladen weight.   A large proportion of the wagons for bulk

5

freight are designed for 25 ton axle loads with access to 80% of the network.

Some of the older track material is still in service on secondary and branch lines but all trunk lines have 113lb./yd. (56kg./m.) F.B. rail welded into continuous lengths, simply seated on pads and held in place by one piece resilient fastenings requiring minimum maintenance. Economics of imported timber have dictated the standard use of pre-stressed concrete sleepers giving greater stability and a life up to 50 years.

Only the highest quality of ballast is used and ballast cleaners enable this to be installed in service to an adequate depth below sleeper to give uniformity of foundation and good drainage.

The track layout on most routes has been simplified with less crossings, many relocated on straight track to give higher speeds and reduced maintenance. Built-up crossings and plated joints have been superseded by cast manganese crossings and welded joints to improve vehicle ride and minimise maintenance. Expensive timber is also giving way to concrete bearers.

A wide range of machines has been developed to carry out relaying and maintenance quickly and to a high standard. Most of this equipment is expensive and requires the possession of at least one track to operate. This is in conflict with the use of the track for commercial business running a 24 hour railway, and developments must be sought in productivity both of machines and track possession hours.

Railways are in competition for labour with modern industry offering superior working conditions and frequently superior remuneration.

The reliance on weekend voluntary workers is being replaced by commitment rostering which includes the weekends in the shift rostering.

The crafts of the track worker are now of a much higher order, emphasising the need for high quality training and re-training.

In 1984 we have therefore a slimmer network in Britain much of it provided with modern durable materials and having at its disposal sophisticated machinery for its maintenance. At the same time it is required to provide a facility which maintains the traditional, but often improved, high standards of safety and enables heavier trains to operate over it at higher speeds. In 1953 the locomotive had the heaviest axle load of the train. In 1984 it has usually a much lighter axle load, with 25 ton axle loads on many freight vehicles.

Looking back over 30 years it is difficult to produce strictly comparable statistics which would assure us that all these changes have been accomplished at a cost which reflects the challenge of productivity of man and machine and provides the higher quality at an acceptable cost.

Despite the difficulties some figures can be produced which are a measure of the change.

1. <u>THE COMMERCIAL PRODUCT</u>

   a) Route miles available for speeds 160k.p.h. or over
      1953   Nil
      1983   1,250 route miles (2,000kms.)

   b) Passenger trains each day with average end to end
      speeds, including stops, over 130k.p.h.
      1953   Nil
      1983   90

   c) Route miles available for freight axle weights up to
      25 tons
      1953   Nil
      1983   8,800 route miles (12,150kms.)

2. <u>THE NETWORK</u>

			1953 miles	1953 kms.	1983 miles	1983 kms.
A	PRIMARY TRACK MILES		10,200	16,415	5,537	8,912
B	SECONDARY " "		10,300	16,576	4,064	6,340
C	TERTIARY " "		8,450	13,599	7,073	11,383
D	RURAL & FREIGHT " "		6,550	10,541	3,806	6,125
		TOTAL	35,500	57,131	20,480	32,960

3. <u>MAIN LINE TRACK RENEWAL</u> (AT 1983 PRICES)

   1953   109 F.B. (54kg./m.) rail, jointed track, softwood
          sleepers, surface reballast only.
          Rail life   15/20 years
          Sleeper life variable 15/30 years
              Cost £140,000 per mile

   1983   113 F.B. (56kg./m.) rail, welded, prestressed
          concrete sleepers, 15 inches (300mm.) new ballast.
          Rail life   20/25 years
          Sleeper life   40/50 years
              Cost £254,000 per mile including ballasting

4. <u>USE OF CONTINUOUSLY WELDED RAIL</u>

   1953   Negligible welded rail

   1983   11,000 (17,700kms.) track miles of welded rail
          2½ million welds, flash butt and thermit

5. <u>MECHANISED PLANT</u>

   Capital value 1953   £400,000/1000 track miles
                 1983   £11,500,000/1000 track miles
   Early tampers each replaced 60 men.
   Modern tampers, by the same measure, carry out the work
   of 120 men.
   Switch and crossing tampers each replace 30 men.
   As a financial measure a Ballast Regulator provides a
   37% return on investment.

The new Pneumatic Ballast Injection Machine will, when in production, probably replace 2½ conventional tampers.

6. MANPOWER

Appendix B illustrates the reduction in manpower numbers and shows the apportionment of savings to system, measured work and incentive payments (P.B.R.), welded rail and use of mechanical plant.   Also shown is the addition required to meet improvements in conditions of employment in line with national industrial trends.

7. MANAGEMENT AND TECHNICAL

a) Maintenance
   1953   Planning for small length gangs by
           supervisors.
   1963/70 Activity measurement and payment by results.
           Introduction of large mobile gangs.
   1973   Measured Daywork Planning.   No incentive
           payments.
   1983   CAMPS. Computer Assisted Maintenance
           Planning System.

b) Surveying
   1953   Theodolite, tape, chain and manual plotting.
   1983   Total station electronic distance measurement
           (E.D.M.) with data recorder, computer
           analysis and automatic plotting.

c) Quality measurement
   1953   By eye and gauge assisted by rudimentary
           Hallade Track Recording.   Visual inspection
           for rail flaws.
   1983   High Speed Track Recording Coach for track
           quality.
           Rail inspection by Ultrasonic Rail Flaw
           Detection Coach augmented by manual ultra-
           sonic equipment.

d) Training
   1953   Little formal training by management.
           Voluntary classes and examinations run by
           the Permanent Way Institution.
   1968   Three formal stages of training introduced.
   1982   Full formal training and skill testing.

CONCLUSION

There are many other statistics which indicate the technical progress over the past 30 years.   Change has taken place - rapid change - but it is not possible to determine beyond debate whether change has in fact been adequate.

The past 30 years show a considerable technological and management effort to meet the commercial challenge of quality, safety and high productivity.

Engineers are part of business management and, whilst seeking always to improve quality both in first cost and total life cost, they can only do so if they are conscious of the harsh competitive climate and provide precisely the standard necessary to meet the specified traffic loads at the required speed.

Unfortunately few credits are given for past excellence and achievement. The competitive challenge is formidable and I hope this conference will be a spring board for a further leap forward.

APPENDIX A

COMPARATIVE ROAD AND RAIL INFRASTRUCTURE COSTS
ROUTE: LONDON - EDINBURGH. 400 miles (640kms.)

1. COACH
   Assumptions 200 journeys per annum.
                65% load factor.   35 passengers.
   Coach licence cost £67.50 per annum.
   Cost per journey £0.34.
   Fuel tax cost per journey 62.8p. x 40 gallons = £25.12.
   Total infrastructure contribution per journey = £25.46.
   Cost per passenger: £0.73.

2. 38 TON LORRY
   Assumptions 200 journeys per annum.
   Payload 26 tons.
   Lorry licence cost £2,590 per annum.
   Cost per trip £12.95.
   Fuel tax per trip 62.8p. x 66 gallons   = £41.45.
   Total infrastructure contribution per trip = £54.40.
   Cost per ton: £2.09.

3. PRIVATE CAR
   Assumed 1.5 passengers per car.
   Total infrastructure cost per passenger journey £8.

4. INTER CITY RAIL PASSENGER
   Assumption 45% loading, 168 passengers.
   Infrastructure charge per train mile is £2.20.
   Total infrastructure charge per journey £880.
   Cost per passenger: £5.24.

5. FREIGHT
   Total infrastructure charge: £115m.
   Total freight carried: 141m. tons.
   Average haul: 71 miles (114kms.).
   Infrastructure charge per ton mile: £0.0115.
   Cost per freight ton £4.60.

These costs are not, of course, real costs.

In the case of rail they reflect the present financial targets to meet sector business objectives and the accountancy conventions for apportionment of infrastructure costs.

In the case of road, they reflect the results of government policy on taxation and apportionment of costs between different forms of road transport, a substantial part of the burden being borne by the private motorist.

Nevertheless, in the competitive area of rail versus coach and lorry, these are the prices which have to be met in commercial rate fixing.

APPENDIX B

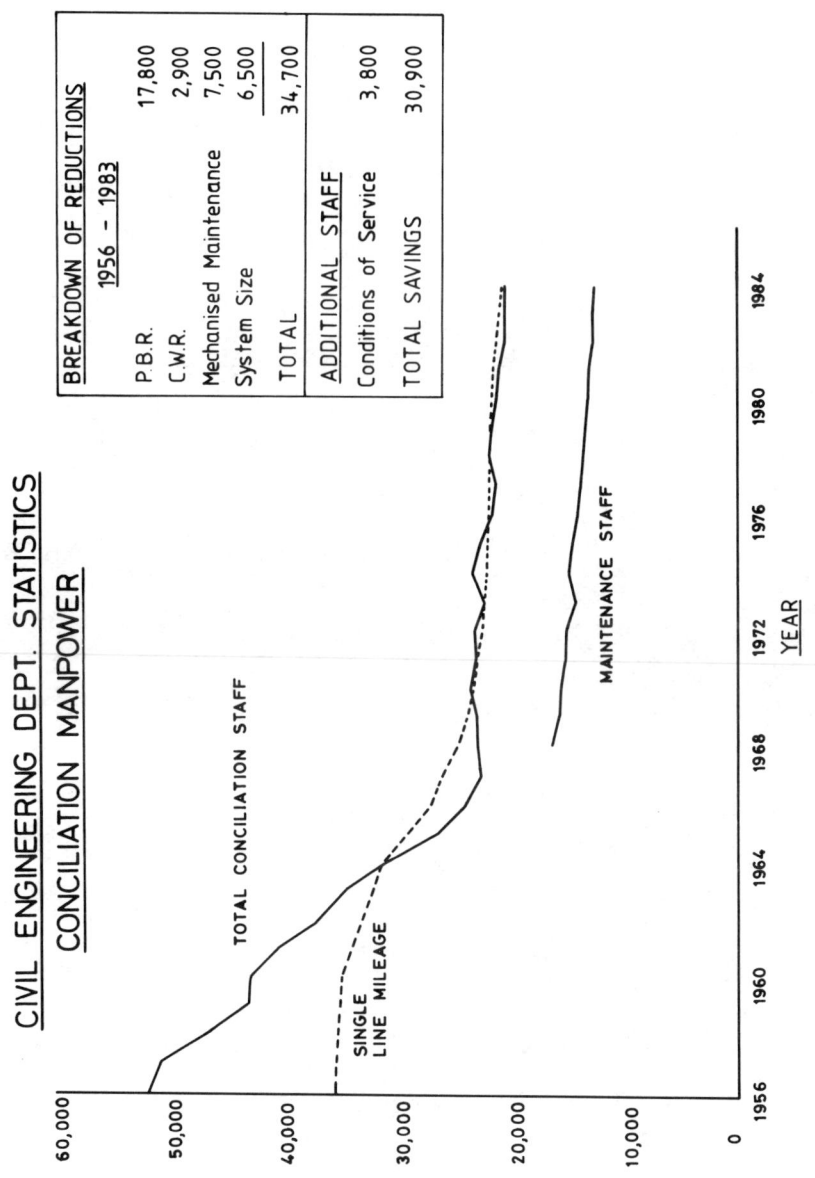

CIVIL ENGINEERING DEPT. STATISTICS

CONCILIATION MANPOWER

BREAKDOWN OF REDUCTIONS
1956 - 1983

P.B.R.	17,800
C.W.R.	2,900
Mechanised Maintenance	7,500
System Size	6,500
TOTAL	34,700

ADDITIONAL STAFF
| Conditions of Service | 3,800 |
| TOTAL SAVINGS | 30,900 |

TOTAL CONCILIATION STAFF

SINGLE LINE MILEAGE

MAINTENANCE STAFF

YEAR

1956 1960 1964 1968 1972 1976 1980 1984

0 10,000 20,000 30,000 40,000 50,000 60,000

11

2 The impact of track technology on heavy haul operations

W. R. FAHEY, BCE(Melb), MBA(McGill), Manager, Track & Technical, Hamersley Iron Pty Ltd, Australia

SYNOPSIS. In Hamersley, the combination of 30t axle load and high tonnage caused unprecedented wear rates and damage while limiting track access for maintenance. It became evident that the initial maintenance strategies would ultimately constrain track capacity. The perceived solution - maintenance based on prevention - required a better understanding of the mechanics of the wheel/rail interface and led the company to join in a major research programme. The objectives: to solve immediate operating problems and to improve overall cost effectiveness.

INTRODUCTION

1. In Hamersley, technology serves the purpose of improving operating economics. A highly competitive environment, internationally and domestically, imposes the need for effective control of transportation costs. It is now obvious that the application of technology to the major areas of cost, such as track maintenance, is critical in achieving this control.

BACKGROUND

2. Hamersley railway transports iron ore over a standard gauge single track of 388kms joining mines at Tom Price and Paraburdoo with two shiploading points at the port of Dampier in the Pilbara region of North-west Australia. Though the climate is hot and dry, cyclones can bring high winds and torrential rainfalls. Ambient temperatures experienced by the track range from 0° to 50°, with rail temp. in excess of 70°.

3. Trains consist of three 2700kw diesel electric locomotives and up to 210 cars with a 30t axle load. Train length is over 2kms, and gross weight about 26,000 tonnes. On the 100km adverse grade of 0.4% between Paraburdoo and Tom Price, three additional locomotives are used as pushers. Speed is 75km/hr maximum in both directions.

The Track

4. The main line from Dampier to Tom Price was built in 1966 to nominal AREA standard with 59kg/m welded rail, untreated hardwood sleepers at 495mm centres, double shoulder plates and dogspikes. Dimensions of ballast and embankments were minimum.

5. Hamersley is a mining operation, not a railway company. Based on the modest project scope envisaged (perhaps 10mt) and

the knowledge then available, decisions favoured construction time at the expense of initial track standards. Consequently, the 288km to Tom Price was built in less than one year, with suspect material selection, construction practices, and water design. Embankments quickly built up a history of settlement, slippage and even failure.

6. From this unimpressive beginning, the track has undergone continual upgrading, in waterways, embankments, ballast section, sleepers (treated Malaysian), fasteners (Pandrol) and rail (68kg/m). Based on the lessons learned from Dampier to Tom Price, the 100km main line extension to the second mine at Paraburdoo began with high standards and, after some 250mgt of traffic, maintenance has been minimal.

7. The constantly heavy axle loads, rising tonnage and train frequency had two important effects: increasing track degradation and decreasing time for repairs. Together, they implied increasing competition for track time between ore trains and maintenance forces. By the mid 70's, with tonnage at 55mgt/y and expected to go on increasing, it seemed that track maintenance would limit the capacity of the system. The circumstances indicated strategy based on prevention, not cure. Cause and effect, in terms of track/train dynamics and, in particular, the wheel/rail interface, had to be understood. To do so, Hamersley joined Mt Newman Railroad and Broken Hill Proprietary Co Ltd in a major research programme.

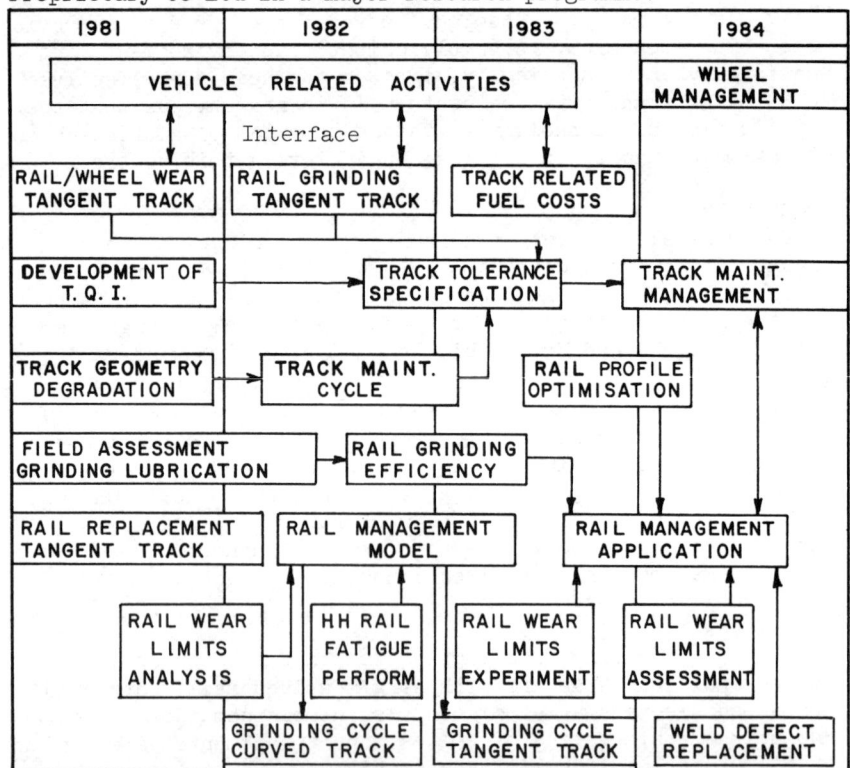

Fig 1. Outline of recent research work

THE RESEARCH PROGRAMME

8. Initially, the programme concentrated on problems in:
. rail performance   . vehicle performance
. track design   . train dynamics
Emphasis now, however, has evolved to analyses of operating
and financial factors in technical/economic models of the
operating system, as illustrated in Fig. 1. Though the interactive nature of the programme is obviously important, further
comment will be restricted to track, with emphasis on those
activities perhaps not generally well known or practised
throughout the industry. In particular:
. rail management   . track tolerances
. sleepers and assemblies   . track protection

Rail Management

9. The purpose is to extract the maximum possible service
life from each rail section, based on technical work in rail
metallurgy, profiling, measurement and wear limits.

10. <u>Rail Metallurgy</u> Findings have been extensively reported
elsewhere (1). Suffice to say here that, on the basis of
costs and benefits as they apply to Hamersley, head hardened
rail appears to produce the best results, though developments
in head hardened alloy rail seem promising.

11. <u>Rail Profiling</u> The concept of reprofiling began through
careful observation of the shapes assumed by rails in service.
In both rails in curves, the crown was seen to develop a slope
encouraging curving behaviour opposite to that desired.
Logically, this effect could be reversed by grinding to move
the point of rolling contact. (See Fig 2). Modelling predicted

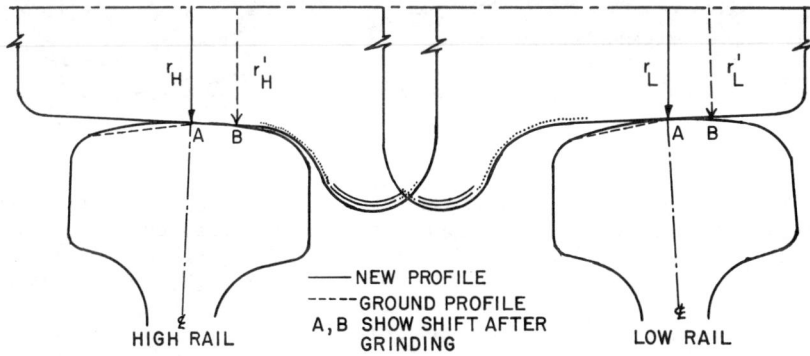

NEW PROFILE
GROUND PROFILE
A,B SHOW SHIFT AFTER
GRINDING
HIGH RAIL   LOW RAIL

Fig 2. Effect of grinding on wheel/rail contact positions.

significant improvement in curving behaviour and wear rates
with appropriate profile shapes. Field tests, before and
after grinding, identified flange force reductions of 50-90%.
Experience has confirmed reduced wear and a weighted average
increase in rail life for all curves approaching 50%. (2)

12. Effect on tangent track may be even more significant.
Prior to profiling, the life of tangent rail was projected at
500mgt. In fact, 68kg/m tangent rail has accumulated about
550mgt to date with a headwear loss of some 10%. Assuming no

unexpected failure mode, the current outlook is for a service
life of 1000mgt or more.

13. Developing an optimal grinding cycle takes into account
the costs of rail grinding, replacement and fatigue defects,
track and vehicle component damage, wheel maintenance and
replacement, derailment risk and traffic disruption. In
detail, planning recognises characteristics which are specific
to the rail in each track section, including curve radius,
rail type, history, current condition, proneness to corrugate
and, critically, the rate of rail degradation. To date, rail
condition has been assessed by inspecting visually, painting
rails to determine contact bands, and measuring discrete cross
sections with a clamp-on gauge. Hopefully, this will change
with the development of a device to measure and record headwear and shape which Hamersley initiated through Australia's
Commonwealth Scientific & Industrial Research Organisation. (3)

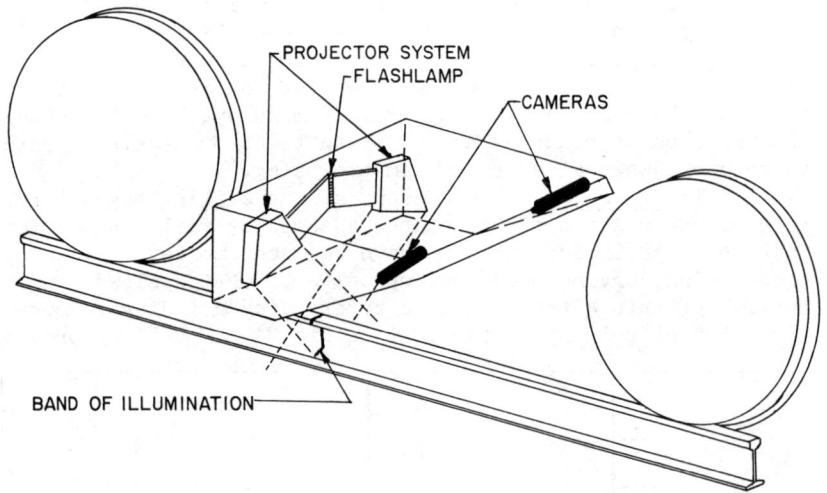

Fig 3. Headwear measuring system

14. As illustrated in Fig 3, a thin sheet of high intensity
light is projected on to each side of the rail head and the
reflected image captured by two solid state, high resolution
cameras. The signals are converted to digital form, the two
half profiles reconstructed from the array, and the image
compared with stored data of an original rail head. After
extensive laboratory development, field testing demonstrated
the feasibility by recording at 80kph and 5m intervals with
initial profile accuracy of 0.5mm. Current expectations for
the operating system include:
. simultaneous non-contact measurement of both rails
. ability to operate day or night in most weather
. measurement in either direction of travel
. retractability for clearance of obstructions
. location identification to 1 metre
15. The device is now in process of commercial development
by Futuris in Perth and will be marketed by Tamper.

16.   Rail Wear Limits   The success of the rail profiling
exercise was instrumental in promoting a reassessment of
policy. Because the grinding activity had the major effect of
retarding the side wear rates of the rail head and improving
its general surface appearance, field personnel demonstrated
increased confidence in extending limits then current. Rail
life crept up to instances of almost double the former limit,
without any explicit change in policy. (It appears the
process was contagious!)

17.   The time had come for a systematic analysis, consisting
of theoretical work, supported by laboratory experiments,
particularly simulating the load regime, and validation tests
on instrumented rails in track. Results indicated that the
critical stresses occurred at the underside of the low rail
head on the field side and were particularly sensitive to the
lateral position of the wheel. The following policy emerged:

Curve Radius(m)	Maximum Allowable Headwear(%)
Up to 500m	35%
500m up to 700m	45%
700m up to 900m	50%
900m and over	55%

Tangent rail headwear has not yet been specified, since current
values are far from any reasonable limit, and the potential
impact of longer term fatigue effects is not known.

18.   Rail management then largely resolves itself into:
optimisation of technical and economic factors in scheduling
the grinding cycle for any given track section; and the use
of current rail conditions, deterioration rates and expected
tonnages to predict rerail requirements far enough in advance
to take advantage of market opportunities.

Sleepers & Assemblies

19.   Careful monitoring of in-track concrete test sections
confirmed the technical performance, increased track
stability, reduction in surfacing, and lower level of track
occupation. Subsequent analysis of the comparative economics
of timber and concrete was decisive in favour of concrete
and the full scale installation programme began in 1980. (4)

20.   After the first year, sleeper movement led to a
comprehensive investigation of the conditions prevailing at the
rail seat. One outcome was a 4mm increase in clip deflection,
to compensate for a degree of relaxation resulting from
normal (and perhaps an element of abnormal) wear and tear
sustained in some 12 years and 550mgt of service by the
existing Pandrol clips. This followed a programme of field
instrumentation and laboratory testing to confirm reserve
capacity. A second outcome was a ribbed pad with variable
stiffness to accommodate inaccuracies in the rail seat planes.
These changes were beneficial, but the experience suggests a
continuing need for investigations into the complex mechanics
of load transfer at the rail seat.

21.   Two other concrete sleeper developments of note are:
. use of a special design at signals where the cables are

introduced through the end of the sleeper for security
during track operations
use of fully concrete sleepered 1:20 turnouts, the first of
which has now carried in excess of 100mgt with a minimal
amount of maintenance. With fully fastened and bonded
swingnose frogs, heelless blades and "self lubricated" plates
this turnout is "state of the art" - for Hamersley, at least.

Track Tolerances

22.   It is easy to accept that wheel and rail wear rates, and
other costs of train operation, are lower when the track
standard is good. It is also a common assumption, however,
that the cost to maintain track in good condition is higher.
The question is therefore whether the benefits of one outweigh
any incremental costs of the other.

23.   The technical work over several years provided Hamersley
with an excellent opportunity to answer this question and, in
particular, determine what track standards would contribute
most to operating economics. Previous investigations of
vertical and lateral loading on the rails, track displacements
and deterioration rates, efficiency of grinding and surfacing
and, in particular, energy consumed and wear created at the
wheel/rail interface, identified the technical relationships
involved. Economic analysis required a variable which would
be descriptive of track standard, and the establishment of
relationships between this variable and operating costs
relevant to the optimisation. The variable chosen is Track
Quality Index (T.Q.I.) which is a composite of standard
deviations derived from measurements (by the 30t axle track
research car) of relative alignments, surfaces, gauge, cross
level and twist. Each of these variables is weighted according
to its influence on the behaviour of ore cars, determined from
computer modelling.

24.   The management model is up and running, with the appropriate costs and financial factors included. The optimal
T.Q.I.'s for specific track sections can now be identified.
As is the case of rail, track management consists of scheduling
those activities under management control - in this case tamping
and lining, essentially - to achieve the T.Q.I. defined as
economic. In general, the analysis in Table 1 shows that
economics will improve with some improvement in T.Q.I.,
particularly on tangent track, even though the Hamersley track
would be described as good by normal industry standards.

Track Protection

25.   Technology devoted to protecting the track from contingency
events include detection of:
. hot boxes   )
. dragging equipment   )   in place
. rising stream water level   )
. wheel flats   )
. unevenly loaded (or overloaded) ore cars)   under
. poorly performing bogies (as indicated)   development
   by uneven or accelerated wheel wear).   )

Table 1. Cost breakdown with varying maintenance practice

Cost Component	Index of Cost	
	Current Practice	Recommended Practice
Rail Replacement	41.83	37.61
Defect	8.24	6.90
Fuel	16.92	13.14
Tamping and Lining	11.18	18.78
Derailment	0.44	0.33
Bearings	0.13	0.10
Wheel Wear	2.68	0.41
Grinding	18.57	12.98
Total	100.00	90.28

INFORMATION

26.   Increasing emphasis on maintenance management required effective information acquisition and processing. Consequently, a comprehensive Track Information System (T.I.S.), consisting of the sub-systems in Table 2, has been developed. Although the development required a large commitment in time and resources, the payoff, in terms of productivity increase and effective management of maintenance activity, has been extremely high. Indeed, without such a system, the programmes described here would hardly be feasible and the benefits of the technological improvements would not be realised.

CONCLUSION

27.   In the case of Hamersley, the economics of ore transport have led to a much higher standard of track than the original, with respect to stability, capacity, wear resistance of components and geometry. In track management terms, the result of the technology described has been a dramatic change in wear rates, material costs, maintenance manhours and overall productivity. The threat of competition between maintenance forces and ore trains for track occupation no longer exists. In the near future, it is expected that track maintenance will consist of grinding, lining and surfacing by high capacity machines programmed by track maintenance management models using current status reports interfaced with existing data bases. Labour will be largely confined to changing components, at a rate which continues to diminish, and dealing with contingencies. At least, this is Hamersley's vision of maintenance in the context of "Track Technology for the Next Decade".

REFERENCES
1. MUTTON R, EPP D, MARICH S. Rail Assessment. Second Inter. Heavy Haul Railway Conference. Colorado Springs. Sept. 1982.
2. LONGSON B, LAMSON S. Development of Rail Profile Grinding at Hamersley Iron. Second International Heavy Haul Railway Conference. Colorado Springs. Sept. 1982.

3. LONGSON B, SMALL G, HEGEDUS Z. Automated Measurement of
Rail Profiles. I.E. Aust. Railway Engineering Symposium,
Melbourne, Aug. 1983.
4. FAHEY W, PERKINS N. Track Upgrading Cost and Benefit,
Third Inter. Rail Sleeper Conference. Brisbane, Sept. 1979

Table 2. Track information system

Permanent Way	Contains track layout, alignment, profile and features by location, e.g. signals, turnouts. Responds to simple enquiries.
Tonnages	Input derived from Operating System data base inc. ore trains, freight, work trains. Accessed by other subsystems.
Speno Rail Rectification System	Grinding location, start - finish, passes, pattern, time worked, travel, repair. Used in planning grinding programme.
Track Component System	Details on rails, welds, joints, sleepers, ballast, railwear, turnout components. Reports current condition, installation and removal dates, accumulated tonnage.
Rerail Prediction System	Uses tonnage and component data bases with decision rules to predict by month and year, and summarise for next 10 years.
Track Defects System	Details of defects found by Sperry detector car, ultrasonic, mag. particle and visual tests. Enquiries by type, location. Input to Rerail Prediction.
Track Geometry System	Input by data tapes from research car. Enquiries on geometry defects by type, location, size, trend. TQI computed by location and compared with previous. Used for lining, surfacing programmes.
Contract Maintenance Reporting System	Reports manhours, machine hours by type, activity codes, locations. Enquiries by manhours, locations, trades, plant number. Lining and surfacing reports on track turnouts.

3 Civil engineering maintenance of high speed railways

K. WATANABE, Director, Railway Technical Research Institute; K. TAKAHARA, Superintendent, NIIGATA Railway Operating Division; Dr. Y. SATO, Chief, Track Laboratory, Railway Technical Research Institute, Japanese National Railways, Tokyo, Japan; T.OHTSUKI, Director, Paris Office, Japanese National Railways, France

SYNOPSIS. Since the Tokaido Shinkansen Line (New High Speed Railways) opened in 1964, its prominent ride comfort and running stability has been maintained for 20 years. Now the Shinkansen Network stretches to around 1,800 kilometers after the series of further openings of the Sanyo, Tohoku and Joetsu Shinkansen Lines. In this paper, our knowledge in respect to the technology of track structure for high speed railway operation is described and what the track structure should be hereafter is also proposed.

TRACK STRUCTURE
Ballasted track and slab track

1. It was one of the basic questions put to the engineers at the construction of the Tokaido Shinkansen, whether or not the conventional structure of ballast-crosstie-rail combination could withstand a commercial operation of trains running at an order of 200 km/h maximum.

2. According to the inquiries to the advanced railways overseas in the 1950's, it was concluded that if the conventional track structure was adopted the maintenance work to cope with track irregularities would be so enormous that Shinkansen would not be a paying business. Though attempts to use the non-ballasted track were tried, the efforts were in vain with little prospect of success both technologically and economically. Then based on the accumulated results of research, the Track Deterioration Theory (ref.1) accounting for the relationship between deterioration and strength of track and maintenance ability was proposed. Based on this theory, the track engineers of JNR decided to adopt the ballasted track and accomplished the project of the Tokaido Shinkansen.

3. The transport volumes after the inauguration exceeded the predicted volumes, necessitating some improvements on the track structure and the track has exhibited its satisfactory performance, even now, passing 38 million tons per year at a maximum train speed of 210km/h.

4. On the other hand, following the opening of the Tokaido Shinkansen, a high growth rate in Japanese economy caused a sharp increase in the transport volumes of JNR and in consequence the demand and supply of manpower for track maintenance became increasingly unbalanced. In view of this situation, development of a manpower-saving track structure began in 1965 with the following targets :
*   construction cost less than double that for the ballasted track ;
*   vertical and horizontal strength and elasticity equal to or

greater than those of the ballasted track ;
* construction speed faster than 200 m/day ; and
* correctability of track irregularities due to deformation of sub-
    structure higher than +-50mm in vertical direction and +-10mm
    in lateral direction.

5. After the examination of various schemes proposed, the slab track
came into being which is a track constructed with rectangular, flat
precast concrete slabs laid on a solid roadbed with a mixed mortar
(CA mortar) layer of cement and asphalt in between, wherein the
horizontal force acting on the slab is borne by concrete projections
integrated with the roadbed and the slab is fixed to the rails by
means of rail fastenings fully adjustable to the track irregularities
in vertical and lateral directions.

6. The Shinkansen Lines completed in 1974 and thereafter following
the Tokaido Shinkansen abound in tunnels and viaducts which were
constructed for the reasons of topography, environmental pollution and
snowfall. Thus slab track, which can save the manpower for track
maintenance has been extensively introduced (ref. 2,3) on such solid
roadbed sections (accounting for 69-93% of the total line length) .

7. The records indicate that the track-laying speed was 200 m/day
(maximum : 400m/day) on the average and the construction cost about
1.1--1.5 times that of the ballasted track. It is certain that the
maintenance cost will turn out far below initially estimated half of
that for the ballasted track (1/5, if materials replacement cost is
excluded) .

Feature of the two types of track

8. With respect to the running performance of vehicles on them and
the behaviours of them under train passege, there are at present
basically no problems with the two types of track.

9. In the initial stage the ballasted track developed a trouble of
mud-pumping phenomenon in the roadbed, but the trouble has been eliminated through insertion of waterproof sheets between ballast and
roadbed and through strict control of roadbed soil. Subsequently, a
mud-pumping happened on account of ballast pulverisation in the elevated track section and this trouble was also liquidated by laying 25-mm-
thick rubber mats under the ballast, which simultaneously contributed
to mitigation of noise.

10. When the slab track was laid in hot weather, the liquid CA mortar
before injection suddenly geleted and lost its fluidity —a so called "
thixotropy". The problem, however, was solved by improving the charge
characteristic of asphalt emulsion and developing a special additive.

11. To cope with changes in the physical properties and freeze-melt
deterioration of CA mortar in the areas of cold climate, a technology
to lower the ice crystal pressure in time of freezing and melting
through entrainment of air bubbles into the structure, thereby improving the heat sensitivity, has been developed with successful application to the Tohoku and Joetsu Shinkansen.

12. For installation of expansion joints at the ends of long rails,
the slab track is so designed that the rail can slide on the rail
fastenings. Under this design there has yet happened no serious trouble
with the slab track.

13. Shinkansen noise on slab track section is 5dB (A) higher than

on ballasted track section in terms of wheel/rail rolling noise.

14.   The technical prospects of reducing the noise level to that on the ballasted track through optimization of the profile, mass, supporting elasticity and acoustic performance of the track slab are very good. The same can be said about reduction of vibration level.

15.   It sometimes happens that snow mass adhering to the underside of vehicles running at the high speed drops onto the track ballast and causes the stones fly away, inflicting damage to the vehicle and other objects.

16.   In heavy-snow districts registering several meters of snow deposit, there is no problem, because the deposits are melted with warm water sprinkled thereon. In other districts, however, the problem is grappled by preventing the snow from flying under train passege through agglomeration of snow particles by water sprinkling, covering the ballast surface, improving the ballast profile or developing a vehicle underside which permits less deposit of snow.

Track structure for high speed railway

17.   The results of high speed tests at an order of 300km/h testify that controlling track irregularities with long and short wavelength was very important as well as the policy of track maintenance for high speed railways of an order of 200 km/h.

18.   Though the characteristics of vertical track irregularities were substantially clarified through simulation models, those of lateral irregularities had not been elucidated due to some effects of lateral movements of rolling stock, gauge and so on. Therefore, tests were carried out in which track irregularities (mainly, alignment and levels) of different magnitudes and wavelength were set on the Shinkansen track and the behaviours of vehicle and track under various speeds were investigated.

19.   The results of the tests revealed the influence of the magnitude and wavelength of track irregularities on the vehicles at different speeds, yielding a required measuring filter for track irregularities. Thus the procedure for inspection and measurement, the tolerance and the effective method for correlation, of track irregularities have been established. On the slab track, a change in the track profile after correction is exceedingly small and in this respect the slab track is superior to the ballasted track.

20.   The results of the high speed tests also confirm that when the condition of the rail top surface is poor, sudden changes in the wheel load occur as the speed rises. An effective countermeasure for this phenomenon will be to measure the irregularities of short wave length on the rail top surface and remove them by grinding.

21.   The applicability of slab track to less solid sections such as embankments is yet to be confirmed. With respect to the impact on the environment, the ballasted track stands at advantage, while with respect to snow, the slab track does so. Concerning the temperature axial force of long rail, more severe design conditions are imposed on the piers for the elevated structure of slab track. The track weight being less heavy, the design load including the body of the structure will be that much lower. In case of tunnel, the slab track can contribute to a smaller cross section of tunnel than the ballasted track. Thus the most important thing is economy. If the traffic volume

is low, the deterioration will be less serious and accordingly the manpower-saving effect of slab track will be that much less conspicuous. Therefore, ballasted track on the embankment is more advantageous than on the elevated structure for economic reasons. If land price is high and, therefore, the elevated structure is adopted, the slab track can be more advantageous.

22. It will be a JNR's policy to construct further new Shinkansen lines that ; the slab track should be adopted in tunnels or on elevated structure (the embankment is not suitable in urban areas where land tends to be utilized intensively) ; the ballasted track should be adopted on the ordinary sections where earth structure is mainly used.

TRACK MAINTENANCE
Basic conception

23. The first question to be answered initially in the track mainte-nance for high speed railways of an order of 200 km/h was whether the correction and repair can catch up with progress in the track irregula-rities or not. The second question was how to cope with such serious track irregularities as calling for slowdown of train operation. The third one was whether trains could derail or not in case of rail breakage.

24. About the first question, initially the quantitative data were so scanty that it was decided upon to measure the track irregularities as often as possible and to establish a work system capable of executing the required repair. For this purpose the Track Inspection Car was operated every 10 days and the maintenance jobs were subcontracted. Experience has revealed that the 10-day inspection period may be extended, but the practice of measuring every 10 days continues so that the results of the executed jobs under contract may be checked imme-diately by the records of the Track Inspection Car.

25. As to the second question, the individual vehicle vibrational acceleration was measured once every day, selecting an arbitrary com-mercial train ; and by imposing a speed restraint at a spot where the acceleration exceeded the tolerable limit, the repair was finished overnight.

26. This procedure of working is based on the following fact.

27. In a high speed test at about 240 km/h of a 4-unit electric car train before the inauguration of the Tokaido Shinkansen, one vehicle happened to start hunting on a well-conditioned track ; in the fifth run of the train, other vehicle also began to hunt at the same spot where that one vehicle did ; and in the seventh run all the vehicles developed the hunting. At this spot a heavy snaking of the track deve-loped as a result of this testing.

28. Thus the above mentioned procedure has been worked out conside-ring that the snaking of the track had been caused as the result of a poor-performance bogie having hunted and thereby exerting a lateral pressure of about 10 t four times ; that a lateral vibratinal accelera-tion of a good-performance bogie has been observed as it developed hunting ; and that the possible number of passeges for a specific vehicle over a specific spot was 1.5 passes /day at the most.

29. At present a different system as described later is in practice based on the built-up data and experience and a vibration measurement of commercial trains is also done supplementally. It goes without

saying that the driver is authorized to take an emergency step of
slowdown by reporting the matter and to have the necessary instruction
issued via train radio and CTC.

30. As for the rail breakage, the third critical problem, a sufficient safety was verified through the experimental train runnings on
20-mm-gap at the speed of 200km/h. As the result of the experiment,
it was regulated that the width of the gap due to rail breakage must
be up to 50mm ; and some necessary expansion joints were installed on
bridges.

31. About periodic patrolling of track or combination of regular
successive maintenance scheme and random, irregular maintenance scheme,
the conception is not different from the general track maintenance.

Control of track irregularities

32. The track irregularities to be controlled on high speed railways
have five items : gauge, cross-level, longitudinal-level, alignment
and twist. Initially, the longitudinal-level and the alignment were
determined just as on the narrow-gauge lines in terms of 10 m chord
versine ; and the tolerance limits were practically correlated respectively to the vehicle vibrational accelerations in the vertical and
in the lateral direction. Theoretically and empirically it has been
verified, however, that these values alone are insufficient for high
speed sections of an order of 200 km/h. Thus instead of 10 m chord
20 m one has come to be measured or calculated for the purpose.

33. If the Shinkansen is to be run faster, in connection with the
vibration of vehicle body and bogie the track irregularities of long
wavelengths will come into question. Therefore, as touched upon
earlier, 48 combinations of different track irregularities of 5-80m in
wavelength were set on a part of the Joetsu Shinkansen and trains were
run at various speeds. Thus the tolerance of track irregularities for
each wavelength were determined with reference to the ride comfort
criteria of Janeway, ISO 2631,etc.

34. In the conventional practice the track irregularities were corrected on the basis of measured vehicle vibrational accelerations. On the
other hand, attainment of specific target values for correction in the
long wavelength range marks a significant step toward faster operation.

35. Still another track irregularity involved in speed-up of the
Shinkansen is one which causes wheel-load changes, which in turn
cause track deterioration due to excessing loading, off-loading and an
increased value of Q/P due to off-loading. This is a track irregularity of short wavelength on the rail top surface. This irregularity
has been indirectly controlled by measuring the axlebox vibrational
accelerations, but hereafter the value of the irregularity itself would
have to be quantitatively controlled.

36. Rail surface irregularities such as bend or corrugation wear will
be corrected by a grinding car. Short wavelength irregularity of
considerable size such as the depression at welded joint will better be
corrected using a special partial irregularity correcting machine.

37. Corrective grinding of rail surface will make one of the important jobs in track maintenance in the future high speed railways.

Track inspection and measurement

38. Initially the working speed of the Track Inspection Car for

Shinkansen had been limited to 160km/h maximum due to the restraint of mechanisms to measure the gauge and the alignment. Therefore, the Track Inspection car was obliged to operate at night so as not to affect the intensive train operation in the day time (nearly 100 trains per day). Later with advent of a non-contact measuring system utilizing the laser beam, its operation at 210 km/h was made possible and the car began to be operated utilizing the headway between commercial trains. This has proved a benefit of increasing the availavble time for night-work and, therefore, improving the productivity of manpower.

39.   The optical measuring system mentioned above works such that the beam which passes through a narrow slit in the floodlights located on the opposite rail falls onto the rail surface and a spot at 14mm below the tread along the resultant reflecting line of rail surface is detected with the light receivers set on both sides.

40.   When a high speed measurement takes place in a snowfall, flying snow hinders transmission of the beam. To be applied in such a case, a measuring system utilizing the eddy current has been developed and is now in practical use on the Tohoku and the Joetsu Shinkansen. This system had a drawback that the worn profile of the rail top affects the result of measurement, but the drawback has now been eliminated.

41.   As pointed out in the above, for future speed-up of the Shinkansen it would be necessary to detect exactly and correct the track irregularities directly related to the ride comfort and wheel-load changes.

42.   HISTIM is a track irregularity inspection device developed to meet this need and it is now mounted on the Track Inspection Car assigned to the Tohoku and Joetsu Shinkansen (ref.4) . HISTIM is capable of finding out the real profile of track through two-time integration of the axlebox vibrational accelerations and picking up track irregularities of whatever wavelength (long or short) as well as of yielding as an output through an appropriate filter the necessary track irregularities for the purpose of improving the ride comfort.

43.   Now, the Track Inspection Car for Shinkansen has three bogies as Fig.1 shows. Though this car is operational at the speed of 210km/h, it seems necessary for the purpose of speed-up mentioned earlier, to increase its operation speed by adopting the two-bogie structure like the ordinary vehicle having a good running characteristic rather than the three-bogie structure. If the two-bogie structure is adopted, the present system of measuring chord versine should be replaced by a new measuring system which is to be improved from HISTIM above mentioned.

Information system for track maintenance administration

44.   Most of the track maintenance jobs for Shinkansen are subcontracted and JNR, specifying the site and kind of job, checks the result of work executed by the contractors and thereafter pays them. The JNR personnel at the track maintenance branch office has been responsible for this checking and payment, while this procedure is computerized (ref. 5) .

45.   Data on the track irregularities collected aboard the Track Inspection Car are registered on magnetic tapes, which are then sent to the central computer where CTC is located to be processed in the same evening ; and next morning the terminal set at the branch office produces as an output, a work instruction, which is sent to the the contractors to specify the jobs to be executed ; a work report in the

Track inspection car

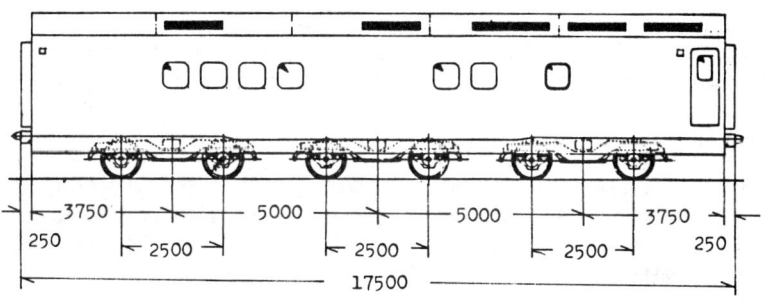

922-11	922-12	922-13	922-14	921-11	922-15	922-16
25000	25000	25000	25000	17500	25000	25000

(a) Train consisting of electric and track inspection cars

3750   5000   5000   3750
250   2500   2500   2500   250
17500

(b) Track inspection car

Scale : mm

Fig. 1 Track inspection car contained in

general inspection train

form of a mark sheet after completion of the work is forwarded from
the terminal set to the central computer, and collated against the
measured records of track irregularities before and after completion
of the work, to formulate a checking document. Thus the sums equal to
the product of unit work price and the qualified length of track i.e.
the checked length of track entered in the document minus the disqua-
lified length of track are paid out to the contractors.
  46. The budget appropriated for the construction of the Tokaido Shinkansen was not ample and accordingly the track structure had to be
designed merely filling the minimum conditions. The actual volumes
carried amounted to about 40 million tons per year against 20 milion
tons per year projected initially for 10-year period following the
inauguration of the line. Nevertheless in the 19-year period the line
successfully transported as much as 610,000 million passenger-km with
safety. What is the secret to the track maintenance performance that
has enabled such a success? Rail weight has been raised from 53.3kg/m
to 60 kg/m, but as a matter of fact what has largely contributed to
the success is the development of the track maintenance administration
system above mentioned.
  47. Another secret lies in the emergency instruction system for train
speed restriction, which works such that when the inspection car

detects any appreciable track irregularity calling for train speed restriction, the warning buzzer sounds and aboard the car the location of the irregularity and other relevant data are numerically printed out. Then the car reports them via telephone to the CTC despatcher, who issues swiftly an instruction for slow down to the trains and an instruction for emergency repair to the maintenance branch office. Such a practice of enforcing the correction of track irregularity where it should be done and enforcing speed restriction where the track irregularity call for it has been firmly established.

48. This is why a scheduled operation of a large number of trains at such a high speed could be accomplished anyhow with minimum capital investment, even though the track structure may not be so robust.

POST-REMARK

49. In discussing the track structure and its maintenance system for high speed railways, optimum relationship must be established among the track deterioration linked to the transport volumes which are associated with the revenue, the structural strength of track which is associated with the capital cost, and the track maintenance volumes which are associated with the operation cost. In this connection it should be stressed that a highly important role is being played by the information system which checks the inspection results of track deterioration (= track irregularities) and verifies the results of track repairs.

REFERENCES

1. INO T. Optimizing maintenance of track and other fixed installations and replacement programmes. Rail International, 1982, 13, 7.
2. MIYAMOTO T., WATANABE K., AOKI M. Development of conventional track. Japanese Railway Engineering, 1975, 6, 3.
3. FUKUDA T. Track structures for the Tohoku and Joetsu Shinkansen. Japanese Railway Engineering, 1982, 22, 1.
4. TAKAHARA K., SATO Y. Recent practices of track management in Shinkansen. Japanese Railway Engineering, 1983, 23, 3.
5. MIYAMOTO T., MOCHINAGA T. Shinkansen track management system. Japanese Railway Engineering, 1976, 16, 3-4.

4 The design, operation and maintenance of a mass transit system

D. J. SHARPE, DIC, MIStructE, MIHE, MHKIE, Engineering Manager & Chief Civil Engineer; J. DRING, MICE, FPWI, Civil Works Manager; B. I. SINGAL, BSc(Eng), IRSE, FPWI, MIPWE (India), Design Manager (Permanent Way) Mass Transit Railway Corporation, Hong Kong

This paper describes the design of permanent way, construction methods, and organisation for maintenance on the Hong Kong Mass Transit Railway. It traces the problems encountered with solutions reached and highlights critical areas. In conclusion, the paper discusses the importance of permanent way and the need for development specifically for mass transit railways.

Introduction
The construction of the Hong Kong Mass Transit Railway (HKMTR) has been described in diverse publications including some papers to this Institution.

2. Construction of the first line, the Modified Initial System (MIS), commenced in November 1975 with contracts

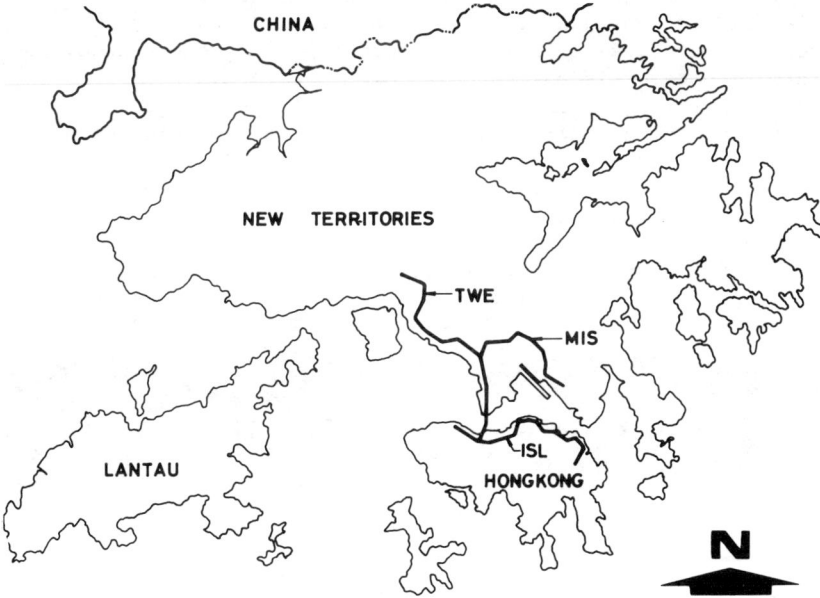

Fig. 1. Hong Kong Mass Transit System

awarded to a number of international and Hong Kong contractors.   Track construction extended over the period of
approximately one year as access became available, and the
full MIS system of 16 km was opened to service in February
1980 - 2 months ahead of schedule.

3.   Subsequently the Corporation has completed the Tsuen
Wan Extension (TWE), and is currently constructing the Island
Line (ISL).   The total system of 38 km of twin track railway
(Fig. 1) with three depots represents a capital investment in
excess of ₤2 billion sterling expended over a period of 11
years, with trackwork costing approximately ₤35 million.

4.   The HKMTR has the usual features of a mass transit
system (MTR) such as sharp curves (up to 300 m radius), steep
gradient (up to 3%), high rates of acceleration and braking,
automatic train operation and protection, and track cir-
cuiting.   Power supply is 1500V D.C. through overhead con-
ductors, with the return current through the running rails.

5.   The distinctive feature of the HKMTR is that it is
intensively used, with approximately 1.2 million people
patronising the system daily.   The trains run at 2 to 2½
minutes headway, opening out to 4 minutes during off-peak
hours.   Dwell time in stations is around 30 seconds.   The
cars are relatively large with axle loads of 16 tonnes and
carry 2500 passengers per 8-car train with 15% seated at a
maximum speed of 80 km/h.

Track Support System

6.   Track on running lines is laid mainly in tunnels, with
some on viaducts and formation.   The track in tunnels and on
viaducts is non-ballasted (NB), for benefits such as:-
reduced tunnel size, weight on viaducts, and maintenance.   On
formation track is ballasted.

7.   NB track consists of rails supported on a resilient pad,
on a continuous concrete plinth made up of two longitudinal
reinforced concrete strips cast insitu on the trackbed (Fig.
2).   Trackbed concrete is cast by the civil engineering contractor with the plinths subsequently cast by the trackwork
contractor.

8.   This division of work gives the trackwork contractor
the opportunity to achieve the specified accuracy in the
finish of the rail seat surface with a minimum amount of

Fig. 2. Non-ballasted Trackbed   Fig. 3. Starter Bars

Fig.4.Plinth Reinforcement   Fig. 5.   Trackbed on Viaduct
concrete to be placed under difficult access conditions.

9.   A shear key in the trackbed and projecting reinforcement starter bars are provided (Fig. 3).   The bars are bent
over by the trackwork contractor to form reinforcement in
the concrete plinths (Fig. 4).

10.   Plinths on viaducts incorporate concrete derailment up-
stands to retain a derailed train (Fig. 5).   Switches and
crossings (S&C) are laid on reinforced concrete slabs.   Rails
are continuously supported except where there are slide baseplates.

11.   At locations sensitive to noise and vibration the trackbed is constructed as a floating slab (Fig. 6) to reduce
transmission of vibrations and associated noise to adjacent
buildings and structures, and is either cast-insitu or
utilises precast concrete units.

Interfaces

12   The plinths and slabs create a physical barrier to cross-
track drainage and cable crossings.   Special provision in the
form of ducts and slots must therefore be made as required.

13.   NB track is low in rail to earth electrical insulation,
therefore reinforcement in the concrete bed is welded to
conduct leakage current back to the substation in order to
prevent electrolytic corrosion of reinforcement and utilities.

14.   HKMTR undertook several full scale and simulated tests
to examine these aspects before finalising reinforcement
details.

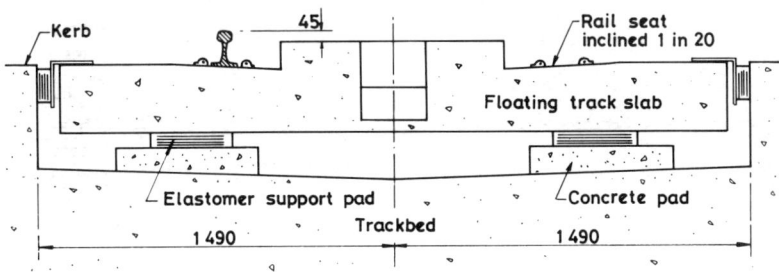

Fig. 6.   Floating trackbed

31

Trackwork

15. The "design and construct" philosophy adopted for the MIS dictated that the rail section was not specified. The successful contractor proposed the use of B.S. 11, 90A rail section which was accepted.

16. Rails are continuously welded on NB track both in plain lines and S&C irrespective of radius. Alumino-thermic welding was adopted since the size of the MIS did not justify the purchase of a Flash Butt Welding plant.

17. Rails are fastened by "Pandrol" clips in shoulders cast or grouted into the concrete (Fig. 7). A continuous rubber bonded cork resilient pad is used between the rail and the concrete to provide track resilience and rail insulation - the latter being supplemented by insulators between the clip and the rail foot.

18. The rail fastening has no facility for vertical or lateral adjustment.

Construction

19. The project programme typically requires the trackwork to be progressed at a rate of approximately 100 m per day. To achieve this the MIS trackwork contractor developed a portable steel shutter system which also locates the shoulders or holding down bolts.

20. Tolerance on the rail seat surface is specified as 1 in 1000 on a random base of 1.5 m length. The rail seat is trowelled to level taking reference from the shutter. The finished rail seat surface is checked immediately the shutters are stripped and any inaccuracies are rectified by mechanical grinding or by building up with an approved resin compound.

21. Rails are received in 18 m lengths and welded into 90 m lengths at the trackwork contractor's worksite for delivery by works trains and then unloaded directly on to the rail seat progressively from this train. The resilient pad is previously glued to the seat. The 90 m lengths are welded on site with special moulds to form continuously welded track.

22. The top edge of S&C slab formwork is used as a reference for finishing the concrete surface to the specified

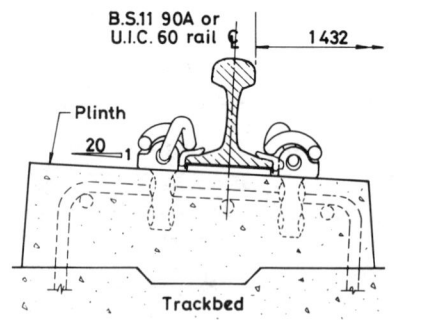

Fig. 7. Track in tunnels

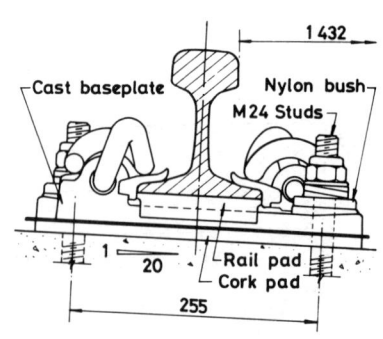

Fig. 8. Track on viaducts

accuracy. Turnouts and special layouts are assembled on the slab and the rails are used as a template to drill holes and grout-in shoulders or holding down bolts.

23. Starter bar positions, accuracy of the shutter and rail fastening locations, acceptance before concreting, and trowelling off are all critical items. Inaccuracy in the first item may require additional starter bars to be grouted in. Inaccurate setting out or trowelling off may require remedial measures or even major reconstruction.

24. The extent of initial problems and the consequential remedial measures reduced to an acceptable level with increasing experience. No problems have been detected on continuously supported rails on the TWE which has now been in service for more than two years.

25. On the TWE viaduct, discretely supported track with baseplates on continuous plinth was adopted. The holding down bolts were cast-in. This appears to have left some bolts in a high state of stress and as a result a number have failed in service.

Maintenance Organisation and Procedures

26. When considering the appropriate structure to adopt for the management of the operational railway it was recognised that, although the system was complex in terms of technical sophistication incorporated into the design, by most standards the size of the System was small.

27. Accordingly, operating and maintenance engineering were combined within one organisation headed by an Executive Director ensuring short lines of communication and effective management control.

28. The Operations Engineering Department is responsible for all maintenance, overhauls and renewal functions and is sub-divided into five groups. The Civil Works Group is responsible for track and trackbed; structures and civil engineering assets such as tunnels, viaducts, stations and depots, and comprises two distinct sub-sections namely: Permanent Way and Structures.

29. The latter sub-section is run on conventional lines utilising a two shift system, covering a normal day shift and a nightshift when work on running lines is carried out.

30. In the case of the Permanent Way sub-section a three shift arrangement has been adopted. The early turn day shift undertakes preventive and corrective maintenance in the depots, and preliminary work for the night shift gangs such as pre-paration of long-welded rails; the late turn provides a small team as insurance against emergency; and the night shift makes up the major part of the labour force.

31. The track is patrolled on foot nightly, with Patrolmen being responsible for the declaration of "Line Clear" messages to the Permanent Way Fault Report Centre. The Centre reports completion of "Line Clear" to the Line Con-troller in Central Control prior to re-energisation of the overhead traction current and the passage of the first

scheduled train.

32. S&C and plain line maintenance gangs working to pre-determined schedules are employed throughout the system during non-traffic hours. Their activities are supplemented by six ultrasonic testing teams using hand-held equipment to check rails and welded joints. There are over 7000 thermit welded joints on the system.

33. Rail transposition has been implemented extensively to prolong rail life, although even so approximately 10% of the total length of running line rails have to be renewed each year. Shop manufactured 90 m long welded rails are transported to site using a simple train adapted from flat wagons.

34. To restrict leakage of traction return current to earth it is essential to keep the track and trackbed both clean and dry. This was recognised, and a gang of cleaners was established for this purpose.

35. To fully utilise these cleaning resources shift rosters are so arranged that on both day and night shifts there is capacity to undertake other cleaning tasks, thereby achieving optimum utilisation of labour.

36. Apart from major civil works all maintenance and renewals are carried out by direct labour. Not only is the cost of contract labour in Hong Kong comparatively high, but expertise in specialist activities such as permanent way is scarce, and high productivity night shift working is more difficult to achieve with contract labour.

Problems Encountered & Solutions Reached

37. Due to the method of casting in shoulders problems with irregular sidecutting occurred on some curves as a result of sub-standard alignment. Replacement shoulders had to be·glued into holes drilled into the plinths. The limited maintenance period dictated that these were located in the space between existing shoulders, with the latter broken out subsequently.

38. Severe general sidecutting occurred within six months of public opening and several of the sharpest curves had to be rerailed on the high side. This occurred before problems with the automatic curve sensitive vehicle mounted lubrication equipment were resolved. Since then, rail life of up to two years and longer has been achieved on the same curves.

39. Extensive rail head corrugation evident before the introduction of rail grinding has been successfully contained, but this requires utilisation of a 24 stone rail grinding train for six nights every week.

40. Gauge corner shelling is the major reason for rerailing. Although reprofiling of both rails and wheels, to attain compatibility, has reduced its incidence, a final solution has yet to be found.

41. There are some sections of the track which are wet. One result has been rapid corrosion of rails in such locations. Sealing or diversion of leaks is a full time activity and, in

addition, at known problem locations, rails and fastenings have been painted to prolong life.

42. Some problems have been encountered with the resilient pad with over compression, permanent set and lateral creep. This condition is apparent during rerailing or transposition. To date, no solution has been found and the pad is renewed during the above activities.

43. In a number of locations the plinths have failed under service. When considering the short time scale of MIS construction it is perhaps not surprising that limited short sections,totallingapproximately 65 m of plinth, have failed. Failures which have occurred are generally the result of poor workmanship. Both precast and insitu concrete have been used for repairs dependent upon location or concentration of plinth damage.

Design Development

44. There were no universally accepted methods for design of concrete bed for NB track. Concern also existed as to the risk of shoulders bursting out of the plinths due to lateral track forces.

45. Full scale tests were conducted with six alternative design arrangements of plinth reinforcement. Hair line cracks appeared in all six designs at lateral forces of between 6 to 8 tonnes, compared with the design force of 10 tonnes. The tests showed that the quantity of reinforcement has little effect on lateral resistance to bursting of shoulders, and that increased strength requires more concrete i.e. wider plinths.

46. Cross drainage through the plinths formed by casting in a 50 mm PVC pipe at 3 metre centres was not effective. The pipes were too small and often ended up being above the trackbed level. In some cases they acted as crack inducers. It was therefore decided to utilise full depth slots for future construction. This change thereby created discrete lengths rather than continuous plinths and also eliminated the possibility of electrical loops being formed by the reinforcement.

47. The lack of track insulation in NB track became apparent at a very early stage in construction of the MIS when the signalling contractor failed to commission track circuits. Investigation showed that this was a surface creepage problem, with the continuous pad forming a barrier such that dirt and moisture were retained.

48. Site and laboratory tests were conducted in an attempt to increase the length of the creep path, and as a result an additional insulating membrane between the rail foot and the resilient pad was installed. This has proved to be an effective solution, and has been adopted as a standard feature for continuously supported rail.

49. On TWE and ISL viaducts an alternative track with discrete supports (Fig. 8) was adopted. Tests showed three

times better insulation than continuously supported rail with
the membrane.

50.   The 90A rail is basically incompatible with the train
wheel profile. Furthermore this rail is not universally
available. To ensure long term availability and improved
rail-wheel compatibility the HKMTR decided to adopt the
international UIC 60 rail section for the ISL.

Critical Areas

51.   MTRs are generally subject to heavy wear, and usually
have sharp curves and steep gradients. It is essential that
dynamic forces between the rail and the wheel are low. This
is achieved by ensuring accurate track construction and
stringent standards of maintenance.

52.   The second critical area is the limited time available
for maintenance, and therefore trackwork should require
minimum maintenance. This is particularly important with NB
track because repairs are complex and difficult.

53.   Some over-design is desirable. In terms of cost this
approach cannot be critical, since trackwork only accounts
for a small percentage of the overall capital cost of a MTR.

54.   Lastly, trackwork has an important interface with the
civil engineering works that precede it and the electrical
and mechanical works that follow it. The requirements of
this interface need to be carefully resolved and incorporated
in the design. Construction usually progresses from one end
of the line. Much of the electrical and mechanical
installation for HKMTR has been carried out from the track,
and therefore the progress of trackwork construction has been
critical to the timely completion of the project.

Future

55.   Mass Transit Systems are being built in increasing
numbers and will continue to be built to reduce traffic
congestion in cities.

56.   A full-scale MTR represents an enormous capital investment and can only be supported in areas of dense population
with a relatively prosperous community. Many metropolitan
cities, particularly in Asia and South America, do not fall
into this category, but have serious urban transportation
problems. Cost reduction must therefore be the most important
consideration.

57.   Trackwork can contribute to this requirement in an
indirect way by considering its importance during construction
and operation. It offers the most efficient means for
installation of services and finishes during construction.
Later, during operation, together with rolling stock, it forms
the backbone of a successful and efficient service. It is
therefore essential to select the most cost effective trackwork design, construction and maintenance methods.

58.   To date track technology developed from traditional
railways has generally been used for MTRs, but it is suggested
that the next decade will see developments in track technology

appropriate to MTRs.

59. Probably the most important technical considerations are the rail and its fastening, however, the most significant must be the low cost of construction and maintenance of the trackbed.

60. Growing experience on MTR track exists. There is an urgent need to pool this experience to establish practical guidelines for track engineers. The HKMTR has played a significant role in setting up a "Fixed Installation Committee" under the auspices of UITP to undertake this task. By such committees and, we hope, papers such as this, successes and failures can be analysed so that we can all learn, for what we do in the next decade may have a major influence on mass transit systems in the next century.

Acknowledgement

The authors wish to thank the Hong Kong Mass Transit Railway Corporation for permission to publish this paper.

5 Track for the next decade—the mixed traffic railway

M. C. PURBRICK, BSc(Eng), FEng, Director of Civil Engineering, British
Railways Board, London, UK

SYNOPSIS. British Railways operates a mixed traffic railway
with Passenger Trains running at 125 mph and Freight Trains
of 25T axle weight at 60 mph. The present policies and
practices of the Civil Engineering Department for maintenance
and renewal of track are discussed. Developments of
materials and methods proposed in the next few years to meet
the requirements of the Business Sectors are outlined.

1. The origins of railways were in freight transport but
they rapidly developed to carry passengers and so became
mixed railways. Most systems since those early days have
carried passengers and freight in various proportions
although a number have retained their specialisation. In
recent times new specialised railways have been built, and
earlier Papers to this Conference have dealt with them.

2. The mixed railway system has, as far as the track is
concerned, similar problems to railways carrying only high
speed passenger trains and railways carrying only heavy freight,
although not necessarily to the extremes. In addition it has
the problems of operating both types of traffic on the same
track.

3. British Railways is a good example of a mixed system. In
1982 60% of B.R.s revenue came from Passengers and 40% from
Freight. In terms of train kilometres Passenger trains were
86% of the total and Freight trains 13%.

4. B.R.s track layout is the legacy of the history of the
development of a multitude of private railway companies in the
19th Century. Very little of the system was designed with high
speeds in mind and much of the route mileage has a high
percentage of curves, which limits the maximum speed. Despite
these limitations B.R. has developed those lines which were
geometrically suitable for 200 Kph operation by the High Speed
Train. These trains have two diesel power units of 2250 HP one
at each end of a train of 7 or 8 vehicles, and more distance is
travelled annually at 200 Kph than any other system using the
existing railway than any other in the world. The power units
of these trains have axle weights of 18 tons and an unsprung
mass of 2.2 tons.

5. Freight operations of B.R. are now nearly all at a maximum speed of 100 Kph with axle weights of 25 tons (25.4 tonnes). B.R. is the only railway in Europe to use 25T axles extensively, most other systems only permit 20T axles on trailing vehicles. Container trains (Freight Liners) with 20T axle weights are permitted to run at 120 Kph.

6. The dynamic load from the high speed power units is greater than that from the 25 Ton axle freight wagons, but of course there are more freight wagons and therefore they have more significance in track deterioration.

7. The bulk of the rail steel used by B.R. is to British Standard 11:1978 Normal Grade. Limited quantities of A Grade and B Grade wear resistant steel together with Austenitic 12/15% Manganese (AM) Steel are also used. A small number of rails made from 110 kg/sq mm 1% Cr. Grade and Austenitic Manganese Steel (12-16% Mn) have been installed. Recently the chemical composition of the latter has been changed to make it more easily weldable to itself. The British Standard is very similar to UIC specification. All the rails are purchased from British Steel Corporation and other than the A.M. steel they are rolled from continuously cast basic oxygen steel; A.M. steel being rolled from ingots teemed from an electric arc furnace. Until recently the finished rail length has been 60ft. but now B.S.C. are supplying 120ft. length rails.

8. Most of the rail bought by B.R. is depot-welded by the Flash-butt Welding Process into lengths up to 1000ft. After conveyance to site by train it is welded again, by the Thermit Skv-F process, to form Continuously Welded Rail Track (CWR). About 10,000 miles of this track now exists on B.R. (i.e. nearly 50% of all track).

9. In the early 1960s, following a programme of analytical and experimental work undertaken by B.R's R. & D. Division in cooperation with the British Steel Corporation (ref. 1,2) B.R. adopted as standard, the section which became BS 110 A. This is identical to the UIC 54 Rail Section.

10. Later work showed that even for 25T axle weights the fatigue life of this section was satisfactory and there was no case for increasing the rail weight on purely rail strength grounds for CWR. However, to cure problems encountered with bolt hole fatigue in the rail web at fishplated joints the web section was thickened up to form the BS 113A section which has since 1969 been B.R's standard rail.

11. This decision has stood the test of time. The critical part of the rail was identified as the upperfillet and to date there have been virtually no failures that could be traced to excessive fatigue stress in this part of the rail.

12. Nevertheless, B.R. has kept its policy under continuous review, particularly in view of the expectation of many European Railway engineers that a heavy rail section would reduce the pressure between the sleeper and the ballast and thus reduce the rate of deterioration of track geometry. We have taken part in the work of ORE Committee D.117 (ref. 3,4) which showed that little difference was found in the geometric

performance which could be attributed to the differing rail
sections. Following the conclusions of the D 117 Committee
therefore, we have so far retained the BS 113A rail as standard
and have achieved reduced sleeper/ballast pressures by reduced
sleeper spacing. Thus although we carry heavier axles than the
rest of Europe we do not intend to follow their trend towards
UIC 60 rail.

13. For all aspects of track maintenance and renewal, B.R. use
a 4 x 4 matrix of speed and weight of traffic for the classification of track. The matrix is shown in Fig.1. Using these
classifications the cumulative rate of rail failures were plotted against age of rail. This is shown in Fig.2.

14. Track Categories A4 and A3 show a very pronounced deterioration trend after being in the track 12 years. In B4 the trend
is less marked and C4 even less so. All other categories
approximate to a linear deterioration with age. It is interesting to note that the worst situation occurs on the high speed
routes A4 and A3 where high quality track is maintained.

15. On the heavy weight slow speed routes D4 a very low rate
of failure exists. Our statistics, at the moment, do not permit
the division of traffic on the A4 and A3 routes between high
speed and heavy axle, but a new recorder is being developed
which will give this information so that further analysis can be
carried out.

16. Whilst the development of defects can be contained by
ultrasonic inspection there is always a possibility of failures
occurring between examinations.

17. The costs incurred as a consequence of unexpected rail
failure are considerable, even without taking account of the
consequences of a passenger train derailment. Equally, if a
rail is found to be defective as a result of ultrasonic
examination, the process of obtaining emergency possession,
cutting in and restressing CWR is very expensive. These
considerations have led B.R. to a decision to renew all rail in
A4, A3 and B4 routes after 12 years service. In the cases of
A4 and B4 this is equivalent to about 250 million gross tons.
The recovered rails are reused in lower category lines. In
order to obtain extended life from the cascaded rails in
secondary lines, on-track planing machines are used to re-
establish an "as new" profile on the head and one side of the
rail. This procedure has produced very considerable savings.

18. On these other lines rails are replaced after specific
amounts of wear have taken place.

19. Rail defects are monitored mainly by a B.R. Ultrasonic
Rail Flaw Detection train which has now been operating for a
considerable number of years. Some six years ago the system
was modernised with an on-board computer to carry out the
primary evaluation. A more detailed evaluation is made on a
computer in the office using records from the car on magnetic
tape. This system was developed to meet the stringent specification laid down for size of flaw to be detected. The main
disadvantage of this system and all other present systems is
the low speed of operation - less than 30 KPH.

		SPEED — MILES/HOUR			
		Over 100	75 to 99	50 to 74	49 or less
Annual Gross Tonnage TONS (MILLIONS)	Over 12	A 4	B 4	C 4	D 4
	5 to 12	A 3	B 3	C 3	D 3
	2 to 5	A 2	B 2	C 2	D 2
	2 or less	A 1	B 1	C 1	D 1

Fig 1 B.R. TRACK CATEGORISATION.

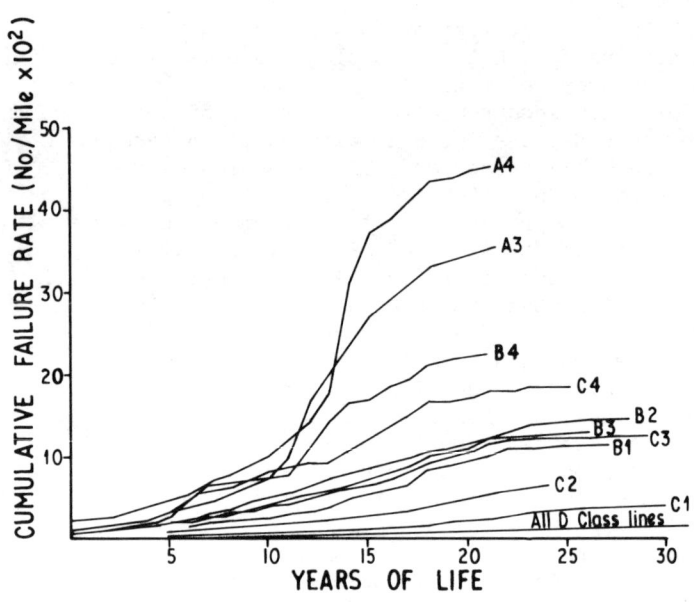

Fig 2. CUMULATIVE RATE OF RAIL FAILURES
PLOTTED AGAINST AGE FOR TRACK
IN VARIOUS CATEGORIES.

20. The present vehicles are now nearing the end of their life and will be replaced by a new unit in the next 2-3 years. B.Rs R. & D. Division is now working on new systems for the new unit and they are confident that speed of operation twice the present can be achieved. This will lead to increased inspections and thus reduce further the possibility of breaks occuring under traffic.

21. Corrugation of the rail surface is a continuing problem and it is contained by grinding. Despite much activity by research workers in many parts of the world, no way has so far been found of preventing its occurrence.

22. The only change in rail in plain track that is likely in the next decade on B.R. is increasing use of Grade A wear resistant steel. Except in circumstances where higher than normal wear occurs, the economic case for more highly wear resistant rail on B.R., as on most other mixed traffic railways is poor.

23. In switches and crossings greater use is made of the high grade steels in places such as some of the London Terminals where intensive suburban services operate on sharp curves. Austenitic manganese steel as rolled rail is used both for switches and fabricated crossings. On high speed lines most crossings are cast AMS and these crossings are also used on very heavily worked freight lines. A big disadvantage of this material is that it is very difficult to weld to BS 11 rail. With normal steel it is the practice to weld through the switches and crossing but since a fishplate joint is necessary with the cast crossing a point of weakness exists. In order to avoid this problem the Research Department at Derby have produced crossing designs using bainitic steel inserts or hearts. This steel has excellent wear properties, is fracture tough and can be welded to BS 11. Trial crossings have now been in the track for some years and have performed well. It is likely that these types of crossing will be increasingly used in forthcoming years.

24. In the 1950s when CWR started to be installed in increasing quantities the concrete monoblock sleeper basically as we know it today, was introduced. The traditional imported softwood timber with baseplates was expensive and relatively short lived. Longer life could be obtained from hardwood but it is higher than softwood in first cost, as were steel sleepers.

25. In addition to the economic advantages of the concrete sleeper, they give better lateral stability.

26. Since the introduction of concrete sleepers on B.R. There have been a number of minor changes in shape, the principal alteration being the section of the sleeper outside the rail where, instead of sloping away on the original design, the top surface was continued level with the rail seat. This change was made to improve lateral stability by making the area in contact with the ballast shoulder bigger. The prestress in the original sleeper was obtainedwith 26 No. tendons of 4.5 mm diameter wire, but this was reduced to 22 No. tendons at the

same time as the change in shape. This resulted from the lessening of the loading following a decision to reduce the standard spacing of the sleepers from 760mm to 703 mm maximum, and 650mm normally, in order to contribute to the requirement for improved lateral stability and reduce the loading on the ballast.

27. The oldest prestressed concrete sleepers have now been in the track for approaching 30 years and have given good service. Whilst there have been a small number of failures of concrete sleepers for a number of reasons, about 7 years ago a pattern of failure was established on certain heavily worked main lines. The failures commenced as a transverse crack through the fastenings under the rail seat, which progressed downwards. Research showed that in these places considerable wheel flat problems existed. The passage of a severe wheel flat was shown to cause concrete sleepers to vibrate in the vertical plane. The third and fifth harmonic of this vertical vibration produces substantial transient hogging rail seat bending moments which results in a tensile stress and causes the cracking.

28. To overcome this problem a further change in design was made, increasing the rail seat hogging resistance moment.

29. The earliest concrete sleepers followed the traditional timber sleeper in being 2590mm in length. It was however appreciated quickly that there were advantages in reducing the length for transport purposes despite the disadvantages of the reduced bearing area and the length was reduced to 2520mm. This has remained as the standard length over the years.

30. Even this length is difficult to handle through single line gantries and with the increasing requirement to carry out renewals without affecting the adjacent track on B.R.s very restricted load and structure gauge, a shorter sleeper is now under development.

31. The critical factor is the bond length outside the rail which with 2520mm sleepers is on the limit for the 4.5mm wire. As an alternative to the 4.5mm wire, increasing use is now being made of 9.3mm strand. With indented strand and the better bond performance that this gives, a sleeper of 2420mm length is possible. The new sleeper also incorporates a number of other changes that we now consider desirable, viz - a broader base to give improved bearing pressure on the ballast, a high centroid of prestress and the upper tendons well up in the section to give good crack control, whilst keeping the weight down to about the same as the existing sleepers.

32. Trial batches of these sleepers have now performed satisfactorily for over two years and the design will be adopted as the new standard which should take us to the end of the next decade.

33. For quite a number of years the timber used in switch and crossings has been Australia hardwood (Jarrah and Kerri), whilst this timber has performed well giving very good life, it is expensive. After trials for a number of years with concrete bearers, B.R. is now using them in increaing numbers.

The extra weight of the bearer is a problem but this can be overcome by the use of modern hydraulic cranes.

34. The key to successful use of concrete bearers is a system for accurately setting out the fastening points so that they can be cast in when the beams are manufactured. This was developed by SRS & SJ in Sweden and is now being tried here. This form of construction will probably become the standard in the future.

35. With the introduction of CWR in the early 1950s the search for the ideal fastening system to go with the concrete sleepers took place. B.R. evaluated 33 different fastenings both by track trials and by laboratory tests. There emerged from this work a list of basic technical and economic requirements that B.R. considered necessary. These were high toe load giving good creep and torsional resistance, consisting of toe load through the expected range of fastening deflection, no baseplate, good fatigue life, no requirement for regular tightening, positive control of gauge, good electrical insulating properties and direct casting of fixings in the concrete sleepers. Two fastenings met these requirements, the Pandrol clip and the Spring Hoop Clip (SHC). Many miles of both types of fastening were installed and performed successfully. The SHC is inserted transversely to the rails and it was not possible to use it with the third rail electrified parts of BR which is about 15% of the system. It is also confined to concrete sleepers. The Pandrol clip is however installed parallel to the rails and can be used in baseplates, on timber, on steel and in switches and crossings. A decision was therefore made in 1963 to standardise on the Pandrol clip. Additional economic advantages came from scale of purchase and reduction in inventory and simplification of concrete sleeper manufacture.

36. Recently Pandrol have produced the 'E' type clip. Whilst basically the same, it has a simplified shape, a smaller diameter bar which provides an increased toe load and fits existing inserts. The shape and bar size makes it a less expensive clip and after 12 months trials B.R. are now adopting it as standard.

37. Closely associated with the clip and concrete sleepers is, of course, the pad. B.R. has used, for the last few years, two types of pad, both 5mm thick. A rubber bonded cork pad for high speed lines and a plastic pad (E.V.A) for heavily worked freight lines - the rubber bonded corks greater resilience being thought more suitable for the vibrations experienced with high speed traffic and the E.V.A.s strength being more suitable for action of heavy freight vehicles. When these two pads were adopted in this way it was an interim decision as it was recognised that there was insufficient knowledge on the characteristics required from a pad and also that the two materials had limitations. It is now becoming clear that a pad thickness more than 5mm is necessary, probably at least 10mm thickness with resilience characteristics between the two materials presently used.

	TRACK SPEED BAND					
	A Over 100 mph		B 75 mph – 99 mph		C 50 mph – 74 mph	
	Top	Line	Top	Line	Top	Line
Quality Band 1. Very Good – No Maintenance required.	1.5 mm or less	0.9 mm or less	1.8 mm or less	1.3 mm or less	2.3 or less	1.9 or less
Quality Band 2. Satisfactory – 90% of route should have S.Ds in this band or QB 1.	1.6 mm to 2.1 mm	1.0 mm to 1.3 mm	1.9 mm to 2.7 mm	1.4 mm to 1.9 mm	2.4 mm to 3.4 mm	2.0 mm to 3.2 mm
Quality Band 3. Poor – Less than 10% of route should be in this band.	2.2 mm to 3.2 mm	1.4 mm to 2.1 mm	2.8 mm to 4.6 mm	2.0 mm to 3.2 mm	3.5 mm to 5.5 mm	3.3 mm to 5.6 mm
Quality Band 4. Very Poor No 200 m should be in this band.	3.3 mm and above	2.2 mm and above	4.7 mm and above	3.3 mm and above	5.6 mm and above	5.7 mm and above

Fig. 3.

38. It is fortunate that there seems to have been a timely breakthrough on pad materials. James Walker have now perfected a process for producing E.V.A. bonded cork which is extremely promising. It is too early to be certain about this material but B.R. is now installing 5000 pads for in-track testing.

39. The inserts in the concrete sleepers into which the Pandrol clips are driven have, in the past, been made from malleable cast iron. Trials with a less expensive pressed steel shoulder have been carried out which have proved successful. The result of all these various points is that the B.R. standard sleeper in the future will therefore probably be 2420mm long, 290mm wide base, have pressed steel shoulders a 10mm thick pad and 'E' type clips.

40. Maintenance of track geometry is a major consideration whatever the type of railway. On a mixed railway the high speed passenger service demands high track geometry standards and the heavy freight causes the deterioration in the track geometry quality. A major advancement has taken place in recent years in that with computers and modern recording cars numerate accurate quality measurements of track geometry can take place. About seven years ago B.R. developed and built its own recording car.

41. This car, which records accurately to +/- 1mm at speeds from 15 KPH to 200 KPH, uses inertial and optical systems i.e. non contact. The data obtained is processed by an on-board computer which feeds a twelve channel analogue recorder and also calculates track quality statistics for every 200 m, which are printed on line by a character printer. This data is also recorded on a magnetic tape for subsequent input into the central computer system. The track statistics are analysed and presented as 'Standard deviation' which describes the average roughness of the 200 mm and as a count of the number of exceedencies of a preset threshold value. The latter is necessary as on otherwise good quality track the significance of one bad isolated defect could be lost.

42. The standard deviation provides a very satisfactory system of quality measurement for vertical level and horizontal alignment ranging from 0 for perfect track to 10.0 for the poorest quality likely to be measured. This form of statistical data of track quality can be related directly to riding characteristics of different vehicles and suspension and also to ride comfort factors for passengers. From these considerations track standards can be set. Clearly they cannot just be arrived at from theoretical calculations alone, and the standards we now use were set after several years of analysis and evaluation of data.

43. For each of the speed bands track standards are set in four Quality Bands, these are shown in Fig. 3.

44. The data analysis from the computer is presented in one form as a bar chart which compares the current data with that from previous runs. This presents an over view for the Regional or Divisional Engineer. The detailed printouts from the car show 200m where maintenance is or will be needed and so the tamping and lining machines which carry out most of our track

Fig. 4

CIVIL ENGINEER'S COMPUTER SYSTEMS STRATEGY

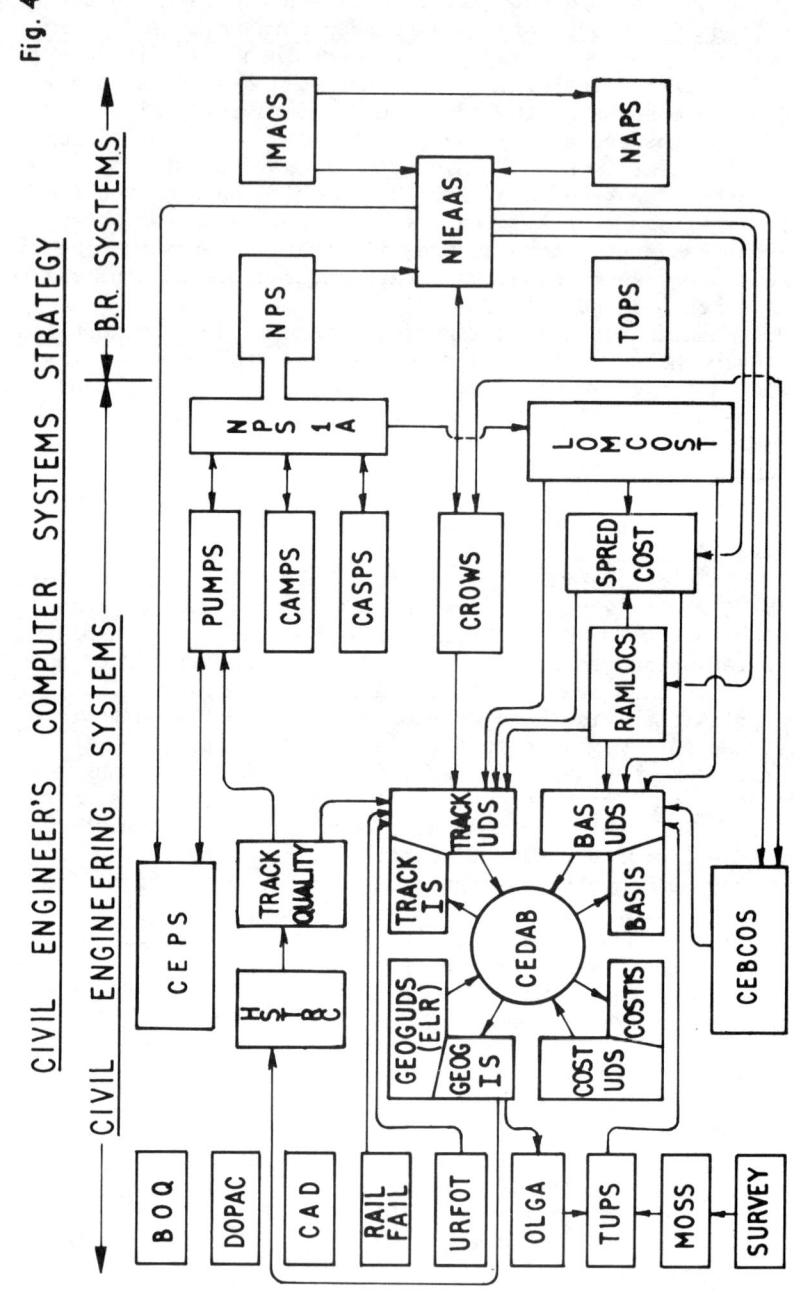

Fig. 5.
CIVIL ENGINEERING COMPUTER SYSTEMS.

BASIS	Bridge and Structures Information System.
BASUDS	Bridge and Structures Up Date System.
BOQ	Bills of Quantities.
CAD	Computer Aided Draughting.
CAMPS	Computer Assisted Maintenance Planning System.
CASPS	Computer Assisted Structures Planning System.
CEBCOS	Civil Engineer's Budget and Costing System.
CEDABS	Civil Engineer's Data Base System.
CEPS	Civil Engineer's Plant System.
COSTIS	Cost Information System.
COSTUDS	Cost Up Date System.
CROWS	Computerised Renewal of Way System.
DEPAC	Drawing Office Planning and Control.
GEOGIS	Geographical Information System.
GEOGUDS	Geographical Up Date System.
HSTRC	High Speed Track Recording Coach.
IMACS	Inventory Management Accounting and Control System.
LOMCOST	Location of Maintenance Costs.
MOSS	Modelling and Survey System.
NAPS	National Accounts Payable System.
NIEAAS	National Integrated Engineering Accounting & Analysis System.
NPS) NPS 1A)	National Payroll System.
OLGA	On-Line Gauging Apparatus.
PUMPS	Planned Utilisation of Maintenance Plant System.
RAILFAIL	Rail Failure reporting system.
RAMLOCS	Resource and Material Locating System.
SPREDCOST	Spread Cost System.
SURVEY	Survey System.
TOPS	Total Operating System.
TRACKIS	Track Information System.
TRACKUDS	Track Up Date System.
TUPS	Tunnel Profiling System.
URFDT	Ultrasonic Rail Flaw Detector Train.

maintenance are programmed only to work on those portions of the track that need attention. Other forms of analysis are also supplied. On the main routes the recording car runs 4 times a year

45. Mentioned above is the exceedence level counts for individual faults. There are two levels one, L2, which requires immediate action by the local gang and L1 which requires action by the local gang within the normal plan of work within the next two or three weeks i.e. before the next machine programme.

46. The knowledge of the track condition that is available from this system at all levels of the organisation and the

consequent control over track maintenance inputs and cost is
one of the most fundamental changes in track maintenance that
has taken place in recent years. More and more reliance and
use of the information will be made in the forthcoming years.

47. With the tamping machine programme as a framework the
work of the maintenance gangs can be formulated into broad
plans three months in advance taking into account seasonal
variations. Detailed work requirements for the gangs are also
noted by the Area Civil Engineer and his supervisors during the
course of their inspections. All of these requirements with a
priority rating are input into a computer system (CAMP)
(Computer Aided Maintenance Planning) which is used to plan
work of each gang.

48. Most major railway systems depend on tamping and lining
machines to carry out the bulk of the maintenance. These
vibrating tamping techniques were developed in order to reduce
the high labour content in the old manual maintenance systems.
Mechanical tamping is the mechanisation of the old 'beater'
system of maintenance. Before mechanisation on B.R. the manual
system used in most circumstances was the measured shovel
packing system involving the measuring of the quantity of
chippings required to be placed under the sleeper to correct any
void under the sleeper and the dip in the longitudinal level of
the rail.

49. The current tamping system requires retreatment over a
cycle of from six months to over 3 years depending upon the
traffic, track material and the nature of the formation and
drainage. As traffic passes over a newly tamped piece of
track deterioration gradually takes place until the track is
approximately in the same state as it was originally.

50. The measured packing system lasted much longer than
machine tamping but was very labour intensive. It was thought
that a mechanised measured packing system would have many
advantages. B.R. has therefore now developed an experimental
prototype machine that does this. The machine on a preliminary
run over the site measures voids and vertical level corrections
and then on the second run lifts the track and pneumatically
injects chippings under the sleeper.

51. The results to date are very promising with a probable
interval between treatments of at least 3 times that of
mechanised tamping. This is a very exciting development and a
joint development contract has now been arranged with Plasser
(GB) Ltd., for the construction of a production prototype.

52. Besides the many technical improvements in track, an
area which must not be left out in the pursuit of efficiency
and reduction in costs is that of the management of all aspects
of track work. The power of the computer here is going to play
an increasing role.

53. The Civil Engineering Department of B.R. has an integrated
computer strategy involving over 20 major systems. (Fig 4 & 5).
All 25 Divisions and 5 Regional H.Qs and Board Headquarters will
be linked together and to main frame computers. The terminals
are all capable of acting as stand alone mini computers for

local use of smaller programmes. Whilst these computer systems can be of tremendous benefit it must be recognised that they bring with them a great demand for discipline. Input must be accurate, or the output is worthless, output must be clearly defined to meet well thought out requirements otherwise it will not be possible to digest the mass of information.

54. The thoughts outlined in this paper on the next decade will probably be only a few of those that will take place. There is no doubt that there is still much scope for new and improved materials, designs, systems and methods that Railway Engineers can pursue in the future years.

REFERENCES

1. BABB A.S. ExperimentalStress Analysis of Rails. Proceedings of Institution of Mechanical Engineers, E.1965 Vol.180, Part 1, No. 41.
2. LOACH J.S. and LINDSAY D. The Implications of Running Heavy Axle Loads on BS 110A Rail. B.R., R. & D.D., Report E.579 December 1965 Unpublished.
3. Study of the changes in track level as a function of the traffic and of the track components. RP 2 ORE Committee D117.
4. Study of the change in track geometry as a function of traffic. Additional results. RP 7 ORE Committee D117.

Discussion on Papers 1–5

MR B. I. SINGAL, Design Manager, Mass Transit Railway
Corporation, Hong Kong

Mr Campbell, in your paper, main line track renewal costs are
shown to be greater now than 30 years ago, but if these costs
are calculated per passenger or per ton, as seems more
appropriate, the picture might be different. Do you wish to
comment on this?

MR I. M. CAMPBELL

1983 track renewal costs may appear, at first sight, greater
than those of 30 years ago. Any simple calculation to compare
these costs on a per passenger mile or per ton mile basis will
not allow for the many changes, outlined in the paper, which
have taken place during the period in question and will not
therefore be representative.

If, however, the renewal cost per year of estimated life
is calculated for 1953 and 1983 then the picture given does
change and becomes brighter. The 1953 renewal cost had a rail
and sleeper life of approximately 20 years giving a renewal
cost per year of estimated life of £7000. The 1983 renewal
cost has a sleeper life of 40 years and a rail life of 20
years. If the cost of installing new rail after 20 years,
£88000 per mile, is added to the renewal cost quoted in the
paper then the renewal cost per year of estimated life, over a
40 year period, becomes £8550.

The main difference between these two values is the
additional cost of the formation work, i.e. 375 mm of
excavation and re-ballasting, now undertaken to provide
adequate support for high speed passenger trains (200 km/h)
and 25 t axle load freight wagons (120 km/h).

PROFESSOR K. RIESSBERGER, Technical University of Graz

In the conference we are discussing the costs of track
maintenance and the impression is given that these costs are

major as indeed they are. However, to make the front line of the battle against maintenance costs clear, some figures should be taken into consideration.

(a) For a standard track in Austria with 15 million gross tons per year, consisting of 54 kg rail, heavy concrete sleepers, sufficient ballast and including sublayer and soil treatment, the cost is approximately 5700 Austrian shillings per metre (AS/m). If the soil treatment, sublayer and rail is subtracted (to concentrate on the rail-supporting system itself) the cost involved is approximately 4100 AS/m.

(b) Cyclic maintenance consisting of levelling, lining, tamping, ballast ploughing and compacting accounts for approximately 40 AS/m and is needed every two years.

(c) Ballast cleaning, needed every 15 years, including all preparation and after-work is approximately 270 AS/m. Taking a lifetime of 45 years (to obtain a pessimistic sum for maintenance) the cost, derived from 22 tamping and three ballast cleaning operations, is approximately 40 AS/m per year, which excludes the cost of grinding and small repairs etc.

(d) If the initial cost of approximately 4100 AS/m and the average maintenance cost of approximately 40 AS/m per year is recalculated to the date of track installation (assuming 6% interest and depreciation and a 3% increase in working cost per year) the cumulative maintenance cost only accounts for approximately 22–25% of the initial cost (as a pessimistic estimate).

This clearly indicates the achievements of the mechanization of trackwork. Machines not only replaced manpower but in the first instance also greatly reduced the maintenance costs. The values may not apply exactly to other administrations, but with a reasonable alteration of percentage (up or down) will lead to the same conclusions.

The results do not take into consideration further savings to be obtained from methods of selected track maintenance, which are being introduced in more and more countries, but they illustrate the gains already achieved and thus provide the preconditions for future developments.

MR J. M. CRUDEN, Canadian Transport Commission, Hull, Quebec

Mr Fahey, in focusing on 'maintenance management', mentions in paragraph 6 several preventive maintenance upgrading items which indicate fundamental changes in design, in particular the change from the American Railway Engineering Association (AREA) standard used in 1966 to the use of elastic fasteners etc.

My interest stems from a continuing study of Canadian derailments, reportable under the Railway Act, from 1970 to

Table 1
Derailments reportable to RTC

Breakdown by assessed cause	1970	71	72	73	74	75	76	77	78	79	80	81	82	83
1 Snow, ice, other, hit by train	18	24	29	12	20	16	24	17	17	24	8	3	16	6
2 Slides, unstable slopes, subsidence	6	9	12	10	19	10	5	7	11	5	5	10	14	5
3 Washouts, floods	6	0	3	2	6	4	6	2	1	11	3	5	4	2
4 Track failure—rail buckle	14	12	14	15	14	10	12	18	20	15	14	16	9	14
5 —rail rollover	6	5	11	9	5	12	11	9	8	6	6	9	18	7
6 —gauge restraint	4	2	6	7	6	9	3	9	8	8	7	10	9	13
7 —broken rail joint	33	36	35	29	51	38	33	28	23	33	35	30	26	23
8 —type unidentified	2	4	1	0	2	2	0	0	0	3	0	3	0	4
9 Track geometry	26	21	28	24	51	39	44	30	20	33	25	25	22	15
10 Turnout component defect	0	2	4	7	3	1	4	7	4	3	5	11	10	7
11 Combination (wheel & rail)	5	2	4	6	2	3	0	1	4	5	4	5	8	7
Total track related	120	117	147	121	179	144	142	128	116	146	112	127	136	103
12 Loose wheels	2	2	0	1	3	3	2	4	2	2	2	2	2	1
13 Broken wheels	16	7	17	18	13	19	9	12	18	19	9	9	10	10
14 Broken axles (incl. D.E. units, burnt)	5	10	7	10	7	7	9	9	15	11	7	4	4	9
15 Journal failures—roller bearings	5	3	9	6	8	9	2	9	6	18	17	21	15	16
16 —friction bearings	39	42	33	28	43	35	36	39	22	42	12	32	14	8
17 —unidentified	2	0	0	0	4	3	8	1	12	1	0	0	0	0
18 Brake gear defective/dragging	0	1	3	1	5	5	3	6	4	6	7	5	4	3
19 Draft gear failure	7	7	12	10	12	9	8	13	12	12	15	7	9	7
20 Misc. rolling stock defects	19	25	25	33	50	25	21	35	20	20	27	23	20	15
Total equipment related	95	97	106	107	145	115	98	128	111	131	96	103	78	69

Table 1 (continued)

Derailments reportable to RTC

Breakdown by assessed cause	1970	71	72	73	74	75	76	77	78	79	80	81	82	83
21 Rule violations	16	28	18	17	25	25	28	15	16	18	34	36	37	23
22 Other employee failure	10	9	22	13	25	18	2	4	12	28	10	22	11	12
23 Train control/marshalling	13	10	18	14	17	12	15	13	12	11	27	12	11	9
24 Vandalism at turnouts	3	4	5	9	14	10	16	10	5	11	6	12	14	2
25 Operational/lading misc. defects	18	9	20	18	24	18	20	14	11	19	16	15	28	11
26 Other	14	14	7	11	16	21	10	7	20	7	16	21	10	8
Total operations related	74	97	106	107	145	115	98	128	111	131	96	103	78	69
27 Cause not determined/reported	0	5	6	4	5	9	11	11	15	0	8	0	2	13
Total derailments	289	293	349	314	450	372	342	330	318	371	325	348	327	250
Deduct track motor car cases	17	25	24	15	23	14	10	10	18	22	22	0	0	0
Total 'train' and other derailments	272	268	325	299	427	358	332	320	300	349	303	348	327	250

date, based largely on a review of reports by others and resulting in the causal breakdown given in Table 1. This was part of a report covering three cases of structural failure of track, in which rail rolled over. Causal areas 4-9 inclusive (Table 1) are clearly such as would be notably reduced by a change to elastic fasteners, with reduced rail excitation. However, where Canadian National Railways have changed to track on concrete sleepers, with Pandrol clips etc., causal areas 12-20 inclusive have also been beneficially affected. The indication here may be that the attenuation obtained, in all amplitudes of rail excitation, has so affected the magnitudes of opposed (rail and wheel) rates of change in momentum as to greatly reduce impacts and all related degradation or failure of track and vehicle components.

While there is an ever-increasing bibliography of research studies, many of great interest, these rarely so relate to the engineering specifics with which they deal as to illustrate premises necessarily underlying such 'maintenance management' philosophy as Mr Fahey describes. I should therefore greatly appreciate hearing more on what engineering observations and conclusions led to the indicated change in rail fastenings: what, for example, were the lessons learned from Dampier to Tom Price?

MR W. R. FAHEY

The standards of construction initially adopted formed the basis of Hamersley's subsequent difficulties with derailments, though there is little doubt that the learning process in controlling trains of the type involved also contributed.

It is suspected that track conditions contributed to 13 derailments in Hamersley's history. Almost all these occurred in the early years of operation and were due to embankment failures rather than component failures. There is no recorded evidence that rail rollover featured as a separate cause, though it may have happened as a result of embankment failure in which cars simply fell over. Rail rollover is unlikely because the fastening system, consisting of double-shoulder plates and three cut spikes (two on the inside) was associated with timber in very good condition during the period under consideration.

The decision to use elastic fastenings was prompted by the high wear and failure rates of the original rails, particularly in curves. At first the intention was to phase in 68 kg/m rail with elastic fastenings as the original 59 kg/m rail needed to be changed. Similarly, 2700 mm sleepers were used to replace the 2400 mm sleepers. As traffic continued to increase at a rapid rate, however, it became apparent that the rate of rail failure would impose an impossible task on the maintenance forces. It is this conclusion which accelerated the total replacement of the rail and the use of elastic fasteners throughout.

The lessons learned from the Dampier to Tom Price experience were that the economics of the trade-off between the capital cost of initial track standards and subsequent annual operating and maintenance costs should be addressed explicitly and that both will depend critically on the nature and volume of the traffic to be carried. As an example, the track from Tom Price to Paraburdoo, completed in 1972, was built to high standards and maintenance has been almost negligible in about 250 million gross tonnes.

DR C. ESVELD, Head of Rail Technology and Quality Control, Nederlandse Spoorwegen, Utrecht

Mr Fahey, rail wear seems to be the decisive factor for rerailing on your network. However, with tonnages from 500 to 1000 million gross tons fatigue problems will also occur. Could you give the number of rail flaws (defects) per 100 km of track when rails are renewed?

MR W. R. FAHEY

Rail wear is the decisive factor so far. Hamersley's policy is to run the rail flaw detector car at monthly intervals in the summer and at two-weekly intervals in the winter. Although the rail has accumulated about 550 million gross tonnes in tangent and some gentle curves, there is no evidence that a pattern of increasing fatigue failure is developing. For example, it was not unexpected to receive a report of no defects requiring attention, as happened recently.

Specifically we have not changed any rails so far because of defects and do not expect the car to detect any more than two or three flaws per 100 km at present. We do not assume that this situation will persist, however, and we have planned to do some accelerated fatigue testing of our rail, particularly in the weld areas.

The fact that we grind regularly to maintain the desired head profile may be a factor in our favour.

MR R. E. HARRIS, Project Director, Davy British Rail International Ltd, London

It seems to me that the mistakes made on the Hamersley mine railway were wholly due to the type and level of expertise available to the company in the original development.

If you were given your time again, Mr Fahey, how would you approach the development, or how would you advise other industrial concerns with similar projects in mind, but with little railway expertise, how they should proceed?

MR W. R. FAHEY

The level of expertise available may have been a contributing
factor, but I do not agree that the 'mistakes' were wholly due
to this. The basic decisions which determined the operating
conditions at Hamersley were made in the context of a number
of marketing, strategic and financial circumstances. They may
appear to be mistakes only with the benefit of hindsight. For
example, the 30 t axle load was chosen to minimize investment
in rolling stock for a given payload. This axle load had only
recently come into use in North America at that time and there
was little or no evidence of its damaging effects until much
more experience had been gained.

Since the railroads of North America, with all their
expertise and long history of operating experience, chose to
impose 30 t axles on American Railway Engineering Association
standard track, it is not surprising that Hamersley should
follow this example. What is now generally known is that
track must be of a high standard to cope with such imposed
loading, particularly where unit trains are concerned.

If I were given my time again, with the benefit of the
knowledge gained in the meantime, I would begin with the
standards that Hamersley now has, or is in the process of
implementing. If I were in that position, but without the
benefit of that knowledge, then I would study in some depth
the operating experience of others who were already doing what
I intended to do. That was possible when Hamersley started
up, because Quebec North Shore and Labrador, in Canada, and
Lamco, in Liberia, and perhaps others, already had such
operations. I am not sure to what extent advantage was taken
of that, nor am I very sure that I would be able to draw firm
conclusions in that particular case because their approaches
have been different. However, the number of projects now in
existence and the common factors which can be deduced from
their operations give clear guidance for any similar project.

If, however, I were now asked to advise on a project
which envisaged a 40 t axle load, then I would say that better
wheels, better rails, stronger bearings and bogies and strong
track of very high geometrical standards would all be
required. The point is, though, that all of this is
extrapolated from current experience and one would get little
guidance from elsewhere. If additional conditions were
imposed in the form of minimum time, resources and money to
start up, then you or I might not be in a good position to do
other than what Hamersley did in the beginning and to hope
that technology caught up quickly.

DR C. ESVELD, Head of Rail Technology and Quality Control,
Nederlandse Spoorwegen, Utrecht

For the correction of vertical weld geometry, Japan National
Railways uses a special milling machine. The welds are not

previously bent upwards as Nederlandse Spoorwegen does at present and is envisaged by various other administrations by using the STRAIT facility. Dr Sato, do you think that it might be beneficial to correct the geometry by bending first?

DR Y. SATO

Yes, I think that it might be beneficial to correct the geometry of a weld deformed badly at the top surface. In practice we have bent many such welds upwards on the Tokaido Shinkansen.

The method is different from your STRAIT method, which is a cold-working procedure. Our method is a hot-working procedure without the use of special machines. The process is as follows: the rail fastening devices near the weld bent upwards are removed; the weld deformed at the top surface is held upwards; the rail is heated; water is poured on and thus the weld is bent upwards; the supports are taken off; finally the rail fastening devices are refastened.

MR F. I. MAU, Civil Design Engineer, BHP Engineering, Sydney

Dr Sato, please could you answer the following questions?

(a) How is equal elasticity achieved (paragraph 4: slab track compared with ballasted track)?

(b) Please provide details of the rail fastening system mentioned in paragraph 5. How is resilience provided? Do you have any comments with respect to slab track on the wear of components, e.g. replacement cycles, or on wheel-rail wear, or of non-suspended parts of vehicles?

(c) What are the spacings of expansion joints? Why are they needed rather than having continuously welded rails (paragraph 12)?

(d) What is done to attenuate noise (paragraph 13)?

(e) Please explain the track loading philosophy discussed in paragraph 35.

(f) What methods of rail defect detection are employed?

(g) What rail types and weld types are used?

(h) Is vibration a problem (paragraph 14) and how is it overcome?

DR Y. SATO

(a) The spring constant for supporting the rail of ballasted track on solid bed is calculated as 69 MN/m by taking that of the tie pad to be 90 MN/m and that of the ballast to be 300 MN/m. Then, the spring constant of the pad for the rail fastening device is determined as 60 MN/m which is smaller than the above value.

(b) The rail fastening device of type 8 mentioned in

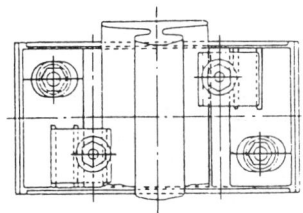

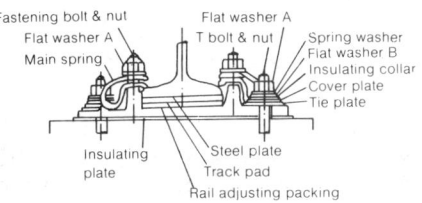

Fig. 1. Model 8 rail fastening device

Paper 15 which is now the standard for slab track is
illustrated in Fig. 1. The resilience is provided by the
track pad. With regard to the life or the replacement of its
parts, please refer to Fig. 3 in Paper 15.

The rail wear depends on the curvature. At a sharp curve
of R 400 at the entrance to the yard, the rail is
replaced after the passage of 17–18 million tons, but on the
tangent the wear is not a factor that determines the life.

The wheel is reprofiled after every 300×10^3 km of
running. The non-suspended parts are checked and replaced at
the same time, if necessary.

(c) The expansion joint is combined with the insulation
joint, i.e. at every 1200 m on the Tokaido and Sanyo
Shinkansens and at every 1500 m on the Tohoku and Joetsu
Shinkansens. This is done to protect the insulating material
and to control the longitudinal force in the rail. The
replacement of rails is very easy because the work is done
using unit sections.

With the use of glued insulation joints, we are
increasing the length of a long welded rail. Even so we shall
use expansion joints to control the longitudinal force due to
long-range substructure or to facilitate the replacement of
rails, for example, on curves.

(d) The noise originating from the track consists of
wheel-rail noise and ground noise. To reduce wheel-rail noise
a fence is effective. For the ground noise the vibration from
the rail which passes through the track into the ground must

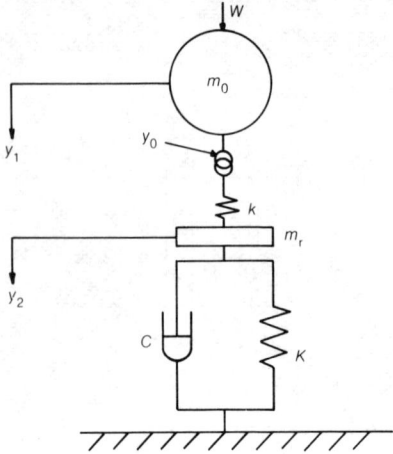

Fig. 2. Loading model of track for an irregularity in the rail top surface: W, load from sprung mass; M_0, unsprung mass; y_0, roughness of rail head surface; k, contact spring coefficient; m_r, estimated rail mass; P, wheel load; K, track spring coefficient; C, track damping coefficient

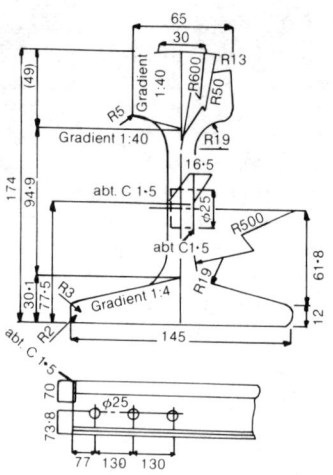

Fig. 3. 60 kg rail (dimensions in millimetres)

be attenuated. To do this, we provide a rubber slab mat under the track slab.

(e) At present, the track loading due to the irregularity of the rail top surface is evaluated with the use of the model shown in Fig. 2 [1]. Here, the masses of the wheel and the effective length of rail and the spring between the wheel and the rail and the spring supporting the rail are important.

(f) The rails of the Shinkansens are inspected with a rail defect inspection car of the ultrasonic type every six months.

(g) The rail used on the Shinkansens is the 60 kg rail shown in Fig. 3. The rails are welded in the workshop by flash or gas welding and in the field by gas or enclosed arc welding.

(h) The vibration is not such a serious problem. To decrease the vibration of frequencies of several tens of Hertz, a mat under the ballast or slab is effective.

REFERENCE
1. Sato, Y. and Kosuge, S. Evaluation of rail head surface configuration viewed from wheel load variation. Q. Rep. 24 (1983) No.2.

DR S. L. GRASSIE, Research and Development Engineer, Pandrol, London

Dr Sato, you refer to a layer of mortar and asphalt below the slab of slab track.

(a) Is it intended that this layer is resilient?

(b) How large are the deflections of the layer?

(c) How thick is the layer?

(d) How durable is the layer?

In paragraph 12 of the paper it is mentioned that there are joints in the long rail on both sides of which the track is free to move.

(a) How long are the sections of rail between the joints?

(b) Over what length on each side of the joint is the track free to move?

(c) How is the clamping force on the rail maintained while permitting longitudinal movement?

On what fraction of the Shinkansen system does corrugation develop? What depth of short wavelength corrugation is permitted before the rail is ground? Typically how long does it take for corrugations to reach this depth (in million gross tonnes or years of traffic)?

Does Japan National Railways have the problem experienced on British Rail of rolling contact fatigue failures of the 'squat' variety? Can these be detected with current inspection systems?

DR Y. SATO

With regard to the CA mortar layer:

(a) it is intended to be resilient

(b) the coefficient of resilience is of the order of 10^3 t/cm per rail fastening

(c) the thickness is a standard 4 or 5 cm depending on the type of slab and the allowable lower and upper limits are 3 or 4 cm and 10 cm respectively

(d) the life is expected to be the same as that of the track.

With regard to the expansion joints:

(a) on the Tokaido Shinkansen, the length of a section is 1200 m and on the Tohoku and Joetsu Shinkansens it is 1500 m

(b) the long welded rail is constrained, with a longitudinal creep resistance of 0.5 t/m per rail; thus the rail movement is limited to the area around approximately 100 m from the expansion joint

(c) the clamping is 1 t per rail fastening.

With regard to corrugations:

(a) on the Shinkansen their development is limited to the braking and driving areas and on steel bridges with long spans

(b) at first the short wave corrugations were ground when they became 0.2 mm deep, and now we have no corrugations that deep

(c) according to our past experience, it took about 70 million gross tonnes (about two years) to attain this value.

We have also experienced 'squat' failures, but we call them 'shelling'. They have been detected by the current inspection system.

MR A. J. S. BLANCHFIELD, Senior Railway Engineer, Freeman Fox & Partners, London

Mr Sharpe, you are correct in saying that a degree of over design in trackwork is desirable but convincing a client of this who has strong ideas on how lean his budget should be can create an incompatibility greater than the mismatch between the 90A rail and the wheel profile. With a consistent resilient support, no possibility of hanging sleepers and a medium axle load, rail strength was not seen to be the dominant design factor. The provision of an additional 30 t of rail steel per track kilometre of which only a part is in the head where it is needed suggests that a special rail section could be developed for use on mass transit railways where the frequency of renewal is a significant factor in track maintenance costs.

Did you not find that mismatch was virtually corrected in the early stages of wear or do you believe that it contributed to excessive wear? On one mass transit railway in Holland the rails have always been installed without inward inclination and thus has created no problems.

It would be interesting to know whether any problems have arisen concerning the rail expansion joints on the viaduct sections of the initial system. The maximum distance between these joints is 250 m and it was calculated that the

differential movement could be as much as 150 mm. Two
structural expansion joints were provided within each station
structure and, although the calculated movement for these was
less than 25 mm, similar rail expansion joints, all of which
were to the conventional scarf design, were provided. Have
the anticipated ranges of movement been realized and has
consideration been given to replacing joints in stations with
plain track?

Concerning corrugations, on the initial system the rails
had corroded heavily during storage before laying. Do you
believe that there could have been a connection between this
and the extensive initial corrugations?

Regarding the fixing of the rail fastenings to the
concrete, do you agree that it is preferable, particularly
where there is no provision for adjustment, for shoulders or
holding-down bolts to be permanently fixed only after the
track has been finally set to line and level?

MR D. J. SHARPE

The decision to use a heavier rail section was based on
operating experience and commercial considerations and had two
basic objectives:

(a) improved service life
(b) assured long-term availability of the rail section since
Hong Kong has no control over standardization of rail sections
in the future.

Development of special rail sections for mass transit
railways is welcome, but their long-term availability must be
assured.

The mismatch between the 90A rail and the wheel profile
was not self-corrected in the early stages of operation. We
believe it contributed to problems such as gauge corner
shelling. When the rails are installed without inward
inclination it is our experience that gauge corner shelling is
worse.

The movement at the rail expansion joints on the viaduct
sections has been less than anticipated and no problems have
been reported. There is possibly a need to replace rail
expansion joints in stations with plane track, but this has
not been given serious consideration.

We do not believe that there is a connection between
corrosion of rails in storage and the extensive initial
corrugation, because rails on the second phase of the Hong
Kong Mass Transit Railway project were adequately protected
against corrosion yet rail corrugation still occurred.

Our experience is that shoulders and holding-down bolts
should be permanently fixed only after the track has been set
to line and level.

MR W. H. HODGSON, Research Manager, British Steel Corporation,
Workington

Rails used in mass transit systems are usually those used for
standard railways. The 54 kg or ARA 115 rails in BS11 A or
AREA carbon steel should be suitable for most modern systems.
The rails are designed for beam strength and stability as well
as head wear resistance. For mass transit with excellent
support characteristics, there may be a case for producing a
rail with an enlarged head to improve wear life only.

An enlarged head could be rolled and produced - the
115 lb rail is often used as a 119 lb heavy head variant.
However, if European rails are treated in this way they become
unbalanced in section, giving rise to cooling problems in
manufacture which in turn gives rise to high initial residual
tensile stresses. Depending on the after-treatment, residual
stresses may have an adverse effect. AREA rails generally
give a better head, web and foot distribution with a resultant
improvement in stress balance.

In addition a significant increase in head weight alters
the neutral axis and gives a completely different stress
pattern from the loaded rail.

On balance it is better to use a well-proportioned rail
with correct stress distribution and to improve wear
resistance by the use of a more wear-resistant material. This
has the added advantage of not requiring a non-standard rail
for future replacement work.

Any mass transit system where premium rails are a
necessity rather than an advantage should look towards
engineering solutions such as track geometry, tyre/rail
profiles, track resilience and bogie design.

MR C. F. BONNETT, Director of Civil Engineering, London
Transport

Although the axle loads and speeds vary greatly it is
interesting to note that the annual tonnages on the most
heavily loaded tracks on British Rail are not very different
from either Australia, Japan or even Hong Kong. In all cases
the most heavily trafficked tracks carry between 30 and 50
million tons in a year. This indicates that the effect of
heavy axles is counteracted by their fewer number when taking
fatigue into account. Mr Purbrick's Fig. 1 shows that over 12
million tons per year even at speeds under 50 m.p.h. enter the
top category.

It is interesting to note that the annual total tonnage
on the London Transport system is about 30 million tons on our
most heavily loaded track.

In view of this I feel that all track components need to
be standardized towards being robust and heavy rather than
light-weight. In most cases a metro will only require a few
components and an economic case could not be made for special

fittings. Because of this I do not agree with your view, Mr Sharpe, that rail fastenings need to be specially designed for metro railways.

DR Y. SATO, Railway Technical Research Institute, Tokyo

For the control of track irregularity, we use the P index value, which is the percentage probability of irregularity above the threshold of 3 mm as shown in Fig. 1. In this case, the following ratings are broadly made for the longitudinal level irregularity in .first and second class lines of narrow gauge where the maximum speed is 120 km/h:

<10	very good
10–20	good
20–30	acceptable
30–40	indifferent
>40	poor

The relation between P value and the standard deviation (σ: mm) is shown in Fig. 2 and the relation between P value and the growth of the irregularity and the annual tamping length ratio is given in Fig. 3.

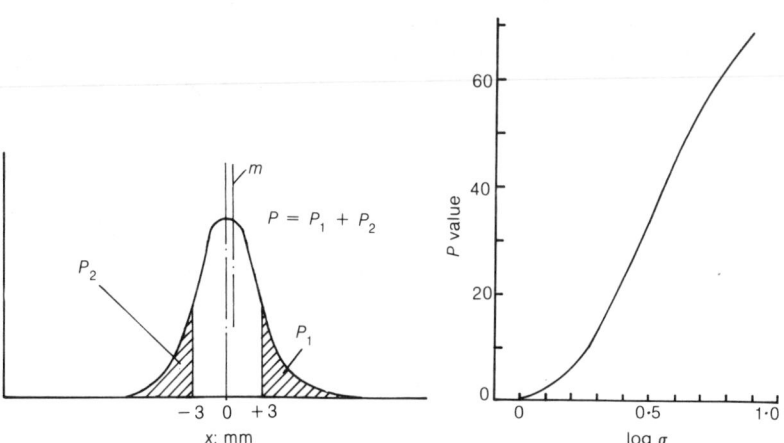

Fig. 1 (left). Track irregularity P value index

Fig. 2 (right). Relation between the standard deviation of track irregularities and the P value

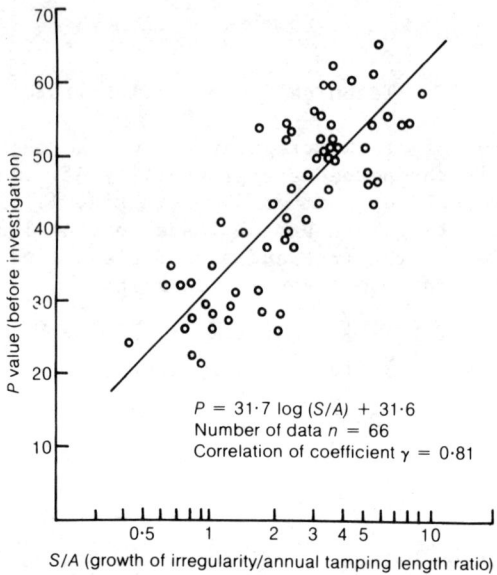

Fig. 3. Relation between P value and S/A in a 500 m length where the track deterioration is balanced with the maintenance work (the change in P value is less than 3)

6 Birmingham Airport Maglev—the development and design of the support structure and guideway

B. H. NORTH, FICE, FIStructE, FIHT, Director, The Henderson Busby Partnership, Ware, UK

SYNOPSIS. The paper gives the background and salient features of the magnetic levitation system followed by a more detailed description of the development and design of the steel guideway and of the pre-stressed and reinforced concrete supporting structure.

INTRODUCTION

1. In 1973 the Research and Development Division of British Rail carried out on behalf of the Department of Environment, a technical and economic comparison between a magnetically levitated (Maglev) and convention wheel/rail suspension systems for both high and low speed passenger transport.

2. The results of the study showed that a Maglev system was feasible using d.c. magnetic attraction for low-speed operation (up to 75 Km/h).

3. In 1975 a test vehicle of 2.7 tonnes was constructed and successfully operated on a 100m length of track at the Railway Technical Centre at Derby.

4. The Client, West Midland County Council decided to provide a Maglev System at Birmingham Airport and to develop and market the system, the People Mover Group was formed. British Rail acted as technical consultants to the Group.

5. Each of the companies forming the Group was responsible for specific aspects of the system as follows:-

The General Electric Co. (GEC)	- project management
	- automatic train control
	- communication systems
	- power supply
	- suspension magnets
Balfour Beatty Power Construction Ltd.	- current collection
	- aluminium reaction rail
Brush Electrical Machines Ltd.	- linear motor
	- propulsion control equipment
Metro-Cammel Ltd.	- vehicle structure
Henderson Busby Partnership	- supporting structure
(as consultants to the Group)	- guideway
	- stations structure

6. The first installation at Birmingham forms part of the Birmingham Airport development which includes a new terminal building. The 600 metres long elevated Maglev system provides a rapid passenger link between this building and the Birmingham International Railway Station and the National Exhibition Centre. A plan and longitudinal section of the system is shown in Fig. 1.

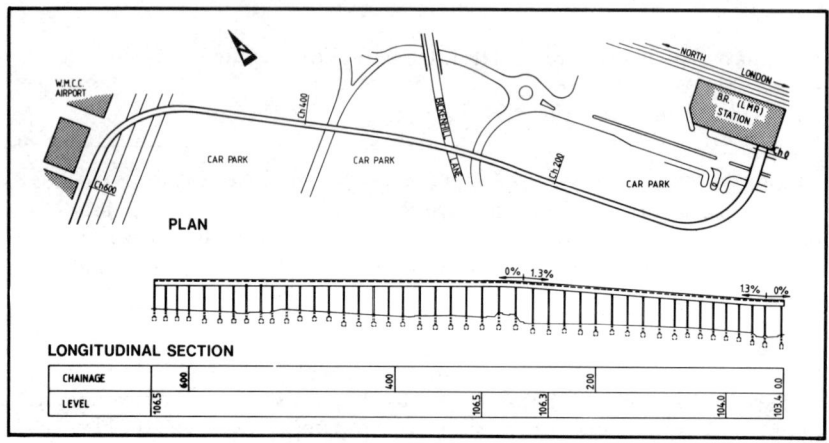

Fig. 1   PLAN AND LONGITUDINAL SECTION

GUIDEWAY DESIGN CONCEPT

1. The Maglev vehicle suspension system uses electro-magnets in lieu of the springs, dampers and wheels and rails used by a conventional vehicle. Since there is no wheel/rail contact and friction by which the vehicles can be propelled, a linear induction motor is used instead.

2. All the loads and forces normally applied to wheeled and tracked vehicles apply in principle to the Maglev guideway. The difference is that, during normal operations there is no contact between the vehicle and its guideway other than for current collection.

3. For reasons beyond the scope of this paper it was decided to provide the vehicle suspension by magnetic attraction rather than repulsion and this obviously affected the configuration of the support system. Basically the vehicles are suspended by 8 electro-magnets, two at each corner, which also provide lateral guidance. Propulsion is by linear induction motor on the vehicle which reacts against an aluminium reaction rail running longitudinally along the centreline of the guideway. Details of the guideway in relation to the vehicle are shown in Fig. 2.

4. The vehicle is mounted by threading it on a removed section of the guideway. Thus it cannot either fall or be blown off the guideway as can conventional wheeled systems. Should there be an electric power failure the vehicle simply de-levitates onto brake pads.

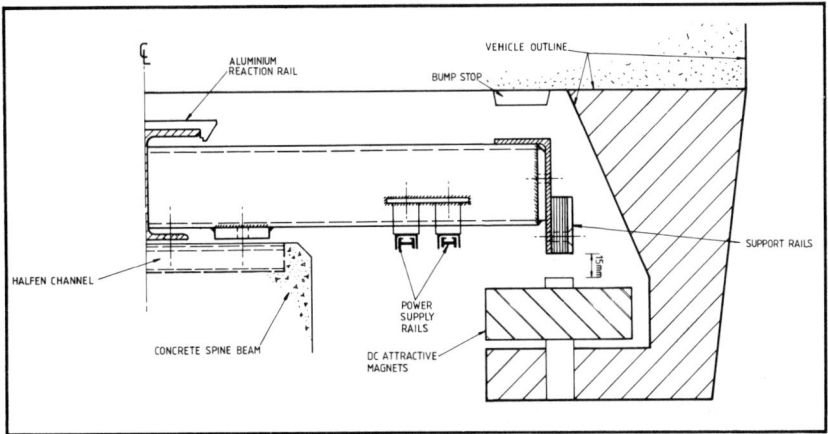

Fig. 2   MAGLEV SYSTEM CONFIGURATION (Not to scale)

5.   There are no moving parts in either the suspension or
propulsion systems. This applies to both vehicle and guideway. Switching will be achieved using moving parts although
not required for the simple shuttle at Birmingham. There is
little or no wear to the various components since there is
only contact between vehicle and guideway when the vehicle
de-levitates when stationary, or due to an emergency breakdown.

6.   The Maglev system is almost immune from rain, snow and
ice since it does not rely on friction for either guidance or
propulsion. Track heating is required on the approach to the
terminal stations to ensure adequate emergency braking in
adverse weather conditions. This is simply achieved by
attaching a heating cable to the supporting angle.

7.   The control system reacts quickly to sudden stimuli such
as wind and is thus well-suited to exposed or elevated sites.

8.   The environmental advantages of the system are excellent
since noise and vibration levels, both inside and outside the
vehicles are very low. There is no air pollution.

9.   The system energy costs for the operation of the system
are similar to those of a comparable wheeled vehicle. The
overall operating costs are expected to be lower when the low
maintenance costs and minimal guideway heating are taken into
account.

THE GUIDEWAY DESIGN PRINCIPLES
1.   The Maglev vehicle has an active suspension system provided by two pairs of transverse flux magnets on each side.
These magnets lift up the vehicle towards the two support
rails fixed to either side of the 'T' shaped track. An automatic longitudinal control system continuously monitors the
position and speed of the vehicle and issues commands to the
vehicle to ensure safe operation. The vertical control system
maintains a nominal air gap between the support rail and magnet of 15mm.

2. Because there is no other form of springing the ride quality depends upon precise control of the gap.

3. The magnetic attraction is provided by direct current which is fundamentally unstable. If left to its own devices the vehicle would either clamp up or fall away. It is necessary to measure continuously the gap and continuously adjust the magnetic force to maintain the gap at the desired value.

4. Thus the levitation control system, which is provided by an on-board computer, continuously monitors the airgaps and the vehicle's vertical accelerations and adjusts the voltages applied to the individual magnets to maintain the nominal airgap while providing an acceptable ride quality.

DESIGN

1. Horizontal Alignment. It was not possible to provide a straight alignment due to several restraints. Firstly, the design of the B.R. station dictated that the Maglev route should be at right angles to the station and access road. The route was then required to avoid land to the south belonging to other owners thus necessitating a sharp right-hand curve. Other restraints, such as the car parking and the entrance to the airport terminal dictated the remaining alignment.

2. The minimum radius specified by the vehicle designers is 40m. Transition curves are based on the cubic parabola and are designed such that the maximum lateral jerk will not exceed 1.25 m/sec^3 at the maximum specified velocity of 13m/s. The maximum lateral and longitudinal acceleration is 0.8m/sec^2.

3. No superelevation is provided due to the fact that the vehicle will be travelling at low velocities on the curves at the entries and egresses to the stations.

4. Vertical Alignment. The maximum gradient specified is 5% although the Maglev System is capable of operating on 10% grades. In fact the actual maximum gradient used is 1.3%.

5. Guideway levels are governed by the platform levels at both stations and by the vertical clearances required over Bickenhill Lane, the access roads and car parks.

6. Whilst these alignment requirements in themselves are not onerous, the design and required tolerances for the vehicle and the design requirements for the structure made the calculations very complex, with accuracy being of prime importance.

7. Tolerances. The tolerances to which the guideway was originally designed and erected, and to which it will subsequently be maintained are as follows:

Level
Absolute   +5, -10mm
Rate of change   1 in 500 on 3m chord

Superelevation   not applicable

Twist
Circular curve and straight   1 in 500 on 3m chord
Transition   1 in 300 absolute maximum

Gauge
Absolute   +5, -5mm
Rate of change   1 in 400 on 3m chord

Alignment
on 10m chord   +4, -4mm
on 20m chord   +6, -6mm
Difference between adjacent readings   +6, -6mm

Offset from Baseline
Absolute   +10, -10mm

8. The two guideways are parallel over the central section
but diverge at both ends to cater for the island platforms.
The vertical alignment is complicated by the requirement that
both guideways be at the same level throughout the entire
length of the alignment.

9. The first step was to quantify the constraints resulting
from the vehicle parameters and to produce graphical horizontal alignments for both tracks embodying these constraints.
From this a rough structure check was undertaken for beam
lengths, pier positions etc.

10. A computer programme was then produced which initially
produced the horizontal centreline alignments and generated
the outer edge of the kinetic envelope. It was then necessary
to extend the program to enable it to produce side and cross
beam lengths and positions, and pier and bearing positions.

11. In order to reduce costs it was decided that all the
main precast beams should be straight, even on the curves.
Thus the distances between reaction rail and edge beams and
the lengths of the cross girders varied through much of the
route length. Each of these lengths was calculated individually using the computer programme.

12. Due to the many changes throughout the development of
the system a close and continuous cooperation was necessary
between the structural and alignment engineers. Despite these
difficulties the design was successfully carried out within a
very short period. An indication of the complexity of the
alignment can best be indicated by the fact that the information provided initially to the steel fabricators of the guideway proved inadequate. The alignment for the centre lines
were then re-run on the computer to produce coordinates at
longitudinal intervals of 200mm.

13. Guideway Details. The support rails are provided by 6
laminations of 100mm x 5mm M.S. strip, bolted together by M16
countersunk bolts at 150mm centres. The reason for using
laminations is to reduce losses due to eddy currents in the
suspension rails.

14. The support rail is bolted to a 200mm x 100mm longitudinal M.S. angle which is in turn bolted to 140mm x 140mm R.H.S.
'sleepers' which are spaced at approximately 1400mm centres.

DESIGN OF THE SUPPORTING STRUCTURE
1. The guideway and its supporting elevated structure
represents as much as 40% of the total cost of the transit

system. There is therefore much to be gained by providing an
efficient structural design. Thus the design development
process not only included consideration of alternative materials such as steel, reinforced and prestressed concrete,
but also special alignment and vibration criteria affecting
the comfort of the ride.

2. During the development of the Maglev vehicles there
were many changes, in vehicle mass, the geometry and to the
levitation and guidance magnet systems, all of which in turn
imposed radical changes in the structural design concept.

3. The original scheme was designed to be constructed
entirely of steel, using simply-supported straight I beams of
12 to 15 metre spans, supported on steel portal frames, this
being the least cost solution. However, this scheme was
rejected in favour of another using steel which was aesthet-
ically more pleasing.

4. This consisted of steel I beams in continuous spans,
curved in plan to follow the alignment, and supported on
V-type piers and pyramid-shaped anchor piers. This design
also proved to be economical.

5. Initial investigations by B.R. into the interaction
between the vehicle levitation control system and the dynamic
response of the supporting structure led to a requirement for
a natural structural frequency of 10 Hz. At this stage in the
development of the vehicle weight was increased from 5500Kg to
9020Kg which was a radical change. It proved impossible to
achieve these requirements with an economic steel superstructure.

6. This led to further detailed investigations by B.R.
which resulted in a reduction of the minimum natural frequency to 8 Hz if the mass of the superstructure was increased.

7. There were also changes in the positions of the struts
of the vehicle resulting in a change from a 'gallows' to
'sleeper' type of guideway support.

8. These fundamental changes resulted in a radically different approach to the structural design and alternatives
using concrete were then considered.

9. Several concrete schemes were tried. The first was
based on using pretensioned simply-supported single box beams,
one for each guideway. The beams were to be curved in plan
where necessary to follow the alignment, reinforced for dead
weight only and made continuous over three spans for live load
by post-tensioning.

10. The gauge of the guideway was then reduced due to changes
in the vehicle design and this resulted in a second scheme
being prepared, similar to the first but using I sections
instead of box girders.

11. Whilst there is no doubt that the under-girder is more
efficient and economic than the through girder system, it
required a greater construction depth and it was this criteria
which finally governed the choice of type of supersturcture.
Thus, the final superstructure to support each guideway
comprised two precast, pretensioned I girders with cast in-

situ post-tensioned cross beams at approximately 1.5 metre centres supporting a central, longitudinal cast in-situ reinforced concrete guideway support beam. (Fig. 3)

12. The main I beams remain straight on the curved sections of alignment and they are spaced further apart to allow for the increase in structure gauge.

13. At the stations solid cast in-situ post-tensioned slabs are used to support the guideway support beam. The main girders support the station and the island platforms in addition to the vehicles.

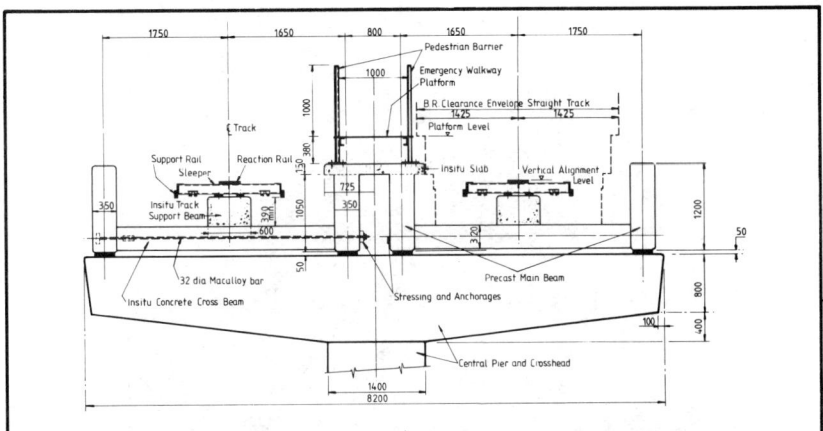

Fig. 3   TYPICAL CROSS SECTION ON STRAIGHT TRACK

14. The most severe design loading occurs when a vehicle suddenly falls from its maximum levitated height of 30mm due to a failure of the levitation system, and is allowed to slide safely to rest.

15. These impact loads are taken on resilient friction mounts on the vehicle providing a simple and reliable emergency braking system.

16. It is unrealistic to apply these large vertical and horizontal impact forces directly to the guideway supporting structure and bearings since their resilience together with that of the mountings absorb much of the energy over a period of time and thus significantly reduce the theoretical instantaneous horizontal and vertical displacements.

17. Analysis of the dynamic response of the vehicle, guideway and structure was carried out by a computer simulation under an impact from two vehicles in tandem on a single guideway.

18. This simulation gave the displacements of the various vehicle and supporting structure components plotted against time, together with the forces. This enabled the rubber bearings, which are particularly sensitive to the braking forces, to be designed.

Fig. 4   THE COMPLETED GUIDEWAY

THE FUTURE

1. The experience gained during the development and design of Maglev leads the author to the opinion that the system may in time supersede wheel/rail systems. Its inherent advantages of a smooth, quiet and pollution-less ride coupled with very low maintenance make Maglev very attractive. Schemes for various airports and ports have already been considered.

2. The second generation of guideway and supporting structure are likely to be simpler, cheaper and of better appearance, mainly due to a better understanding of the vehicle performance and as a result of confidence gained in its operation.

ACKNOWLEDGEMENTS

1. The author wishes to thank the following organisations and people for their help in the preparation of this paper: The West Midland County Council; The People Mover Group; M.K. Czechowski, M.G. Howard and P.G. Porter of Henderson Busby International.

7 A track switch for magnetically levitated vehicles

M. J. LILLEY, BSc, MIEE, Senior Principal Scientific Officer, British Railways Board, Derby, UK

SYNOPSIS. A long-standing problem with magnetically levitated (Maglev) transit systems is designing a switch to transfer the vehicles from one track to another. Alternative configurations have been compared and an active version selected as the optimum solution. A description is given of the design, construction and testing of a prototype switch which has been installed as part of British Rail's Maglev test track.

INTRODUCTION

1. A magnetically levitated (Maglev) transit system offers certain advantages over wheeled alternatives. These arise primarily because a Maglev vehicle has no moving parts and is not in contact with the track during normal operation. The system offers potentially high reliability and a low maintenance requirement. In addition, it is very quiet in operation and can negotiate steep gradients and tight curves.

2. Several Maglev schemes are under development throughout the world. Workers in Japan and Germany are interested principally in high speed systems (up to 400 Km/h), whereas in Britain interest has centred around low speed applications (up to 80 Km/h). The R & D Division of British Rail has designed and built a vehicle and test track at the Railway Technical Centre in Derby, and has been the technical leader in the world's first commercial Maglev link at Birmingham Airport.

3. The B.R. design of track is in the form of a 'T', with levitation magnets mounted on struts which extend below the vehicle and wrap around the track (Fig. 1). This arrangement restrains the vehicle in an emergency and prevents it leaving the track, but poses problems in switching from one track to another.

4. After careful study of all the possible solutions, BR selected a form of active switch as being the most suitable for its application and a development version was built into the test track at Derby. It has proved to be successful and has been cycled through many operations during testing.

DESIGN CRITERIA

5. In designing a switch for Maglev, certain requirements are similar to those encountered in switching a conventional

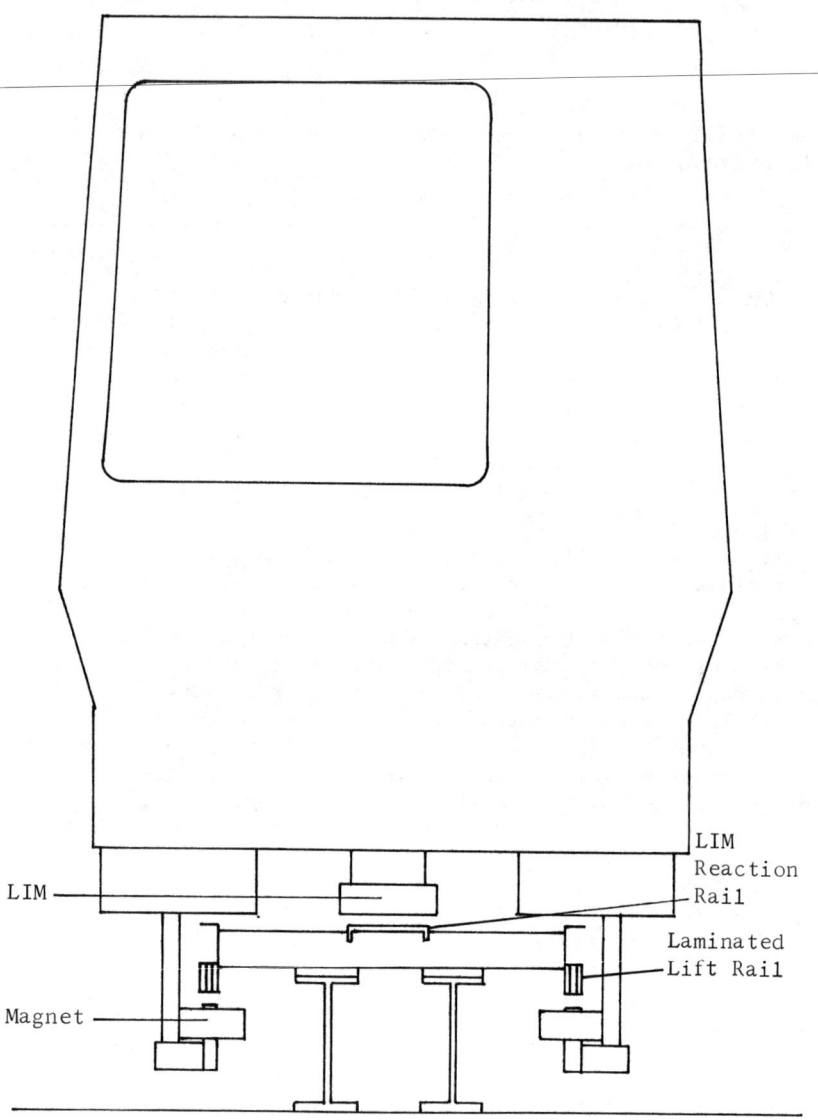

Fig 1. Cross section of Maglev vehicle and track

railway track and some are unique to the Maglev system. The
former category would embrace the following :

a) Operating time and its effect on vehicle headways
b) Cost - installation, operation and maintenance
c) Reliability
d) Means of proving the switch for signalling purposes.

Features unique to Maglev are concerned with providing
continuity of the following track components :

a) Lift rail - mechanical and magnetic
b) Power collector rail to ensure no loss of electrical
   supply which could result in the vehicle de-levitating
c) Reaction rail - electrical and magnetic to eliminate
   thrust variation from the linear induction motor.
d) Upper surface of lift rail for emergency braking

FORMS OF SWITCH
Passive
   6. A passive switch contains no moving parts, and a typical
track arrangement is shown in Fig. 2. Large gaps must be left
in the track to allow the magnet struts to pass through and
therefore special arrangements are necessary to maintain
continuity.
   7. Auxiliary lift rails located outside the main lift rails
are required at all gaps and these extend for some distance
either side of the discontinuity. Additional lift magnets are
provided on each side of the vehicle and by energising the
magnets at one side or the other in combination with the main
lift magnets, the direction to be taken at the switch can be
selected and magnetic levitation maintained across it.
   8. A similar arrangement is required for the power conductor
rails. Additional rails are required at the switch, and
duplicated sets of collectors must be fitted to the vehicle to
ensure no loss of supply which would result in de-levitation.
The system must be designed so that the collectors can run off
the conductor rail at the approach to gaps and re-establish
contact with the conductor rail once the gap is traversed.
   9. Absolute continuity of the linear motor reaction rail is
impossible to achieve without duplication of the motor, which is
unjustifiable on the grounds of cost and weight penalty.
Special arrangements are necessary in the linear motor control
system to ensure the effects of the reaction rail discontinuity
are minimised.
   10. The upper surface of the auxiliary lift rails can be
utilised as an additional emergency braking surface. Extra
skid pads are required on either side of the vehicle, and a
pronounced lead-in must be introduced at the end of each rail to
cater for the drop of the unloaded skid pads on their resilient
mounts.
   11. Since the passive switch itself does not determine the
direction the vehicle takes, proving of the switch in the
conventional sense is not possible and the signalling system
needs to be in contact with the on-board control system in
each vehicle.

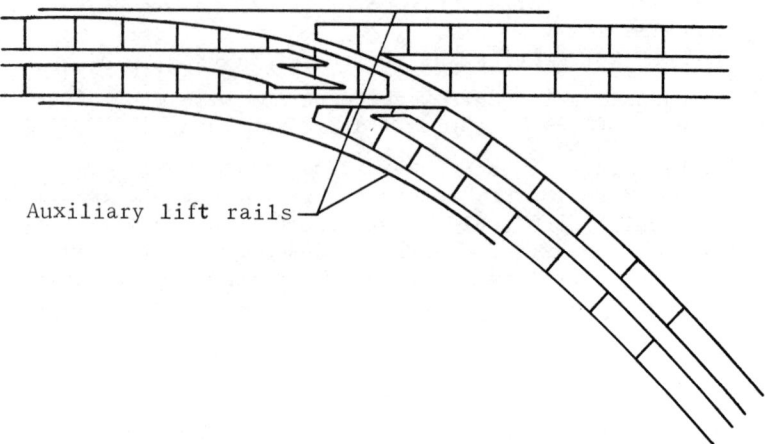

Auxiliary lift rails

Fig 2. Passive Switch

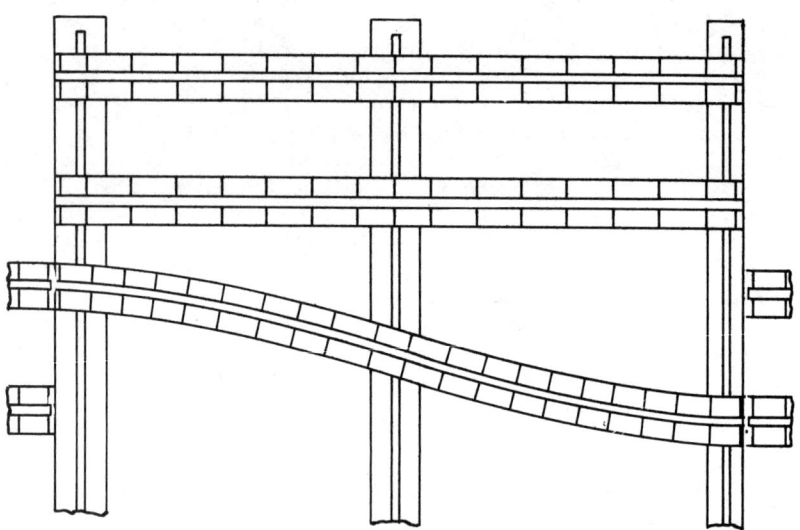

Fig 3. Traverser in crossover position

12. Overall, the advantage of the passive switch is its lack of moving parts, with the associated high reliability and the ability to handle vehicles at short headways since there is no operating time to be considered. The main drawback is the need to duplicate lift magnets, emergency skid pads and electrical power collectors on each vehicle and to provide a more complex on-board control system. The resulting weight penalty is particularly onerous on a magnetically levitated vehicle.

Active

13. General. An active switch works on a principle more akin to conventional railway points, in that a section of track is physically moved to select the route. Continuity of the various elements of the track can be ensured by careful design of the moving sections and these may be locked in position by sliding bolts, which incorporate means of proving the switch to the signalling equipment.

14. Since the track is continuous over the switch, and therefore virtually undistinguishable from normal track, no additional equipment is required on the vehicle.

15. Active switches may take a variety of forms. During the course of British Rail's development work, three configurations were considered in detail and these will be described briefly in the following sections.

16. Traverser. Traversers are familiar pieces of equipment in large railway workshops for moving vehicles from one track to another. Generally, a vehicle is positioned on the traverser and the whole assembly then moves laterally to the required position. The vehicle is then moved off onto the new track.

17. The proposed traverser for Maglev works in a slightly different manner, in its simplest form, the carriage contains not one but two sections of track, one being straight and the other curved. One fixed track leads into one end of the traverser and two tracks lead from the other end, representing the straight ahead and turn-out positions respectively. Fig. 3 shows an extension of this principle to form a crossover between two parallel tracks.

18. In operation, the carriage is moved before the vehicle reaches it so that the route is pre-set and proved in the manner of conventional points. The vehicle can negotiate the traverser at maximum speed in the straight ahead position and its speed through the turn-out is restricted only by the radius of curvature of the curved section.

19. The traverser contains no areas of high technical risk and is expected to be reliable in operation. Its main disadvantage is the cost of manufacturing the large structure required to support the tracks, particularly if elevated above ground level, and the total width of the unit. As an example, a double track crossover as in Fig. 3, based on the track geometry as used at Birmingham Airport where the minimum curve radius is 40 metres, would be 27 metres long and 22 metres wide.

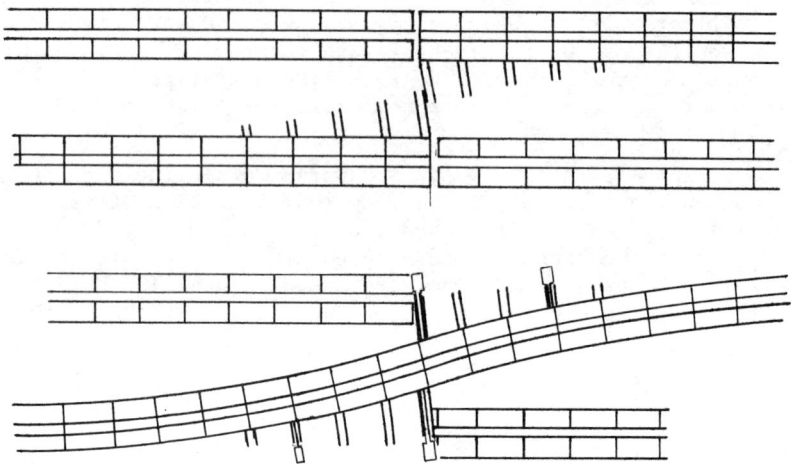

Fig 4. Articulated switch

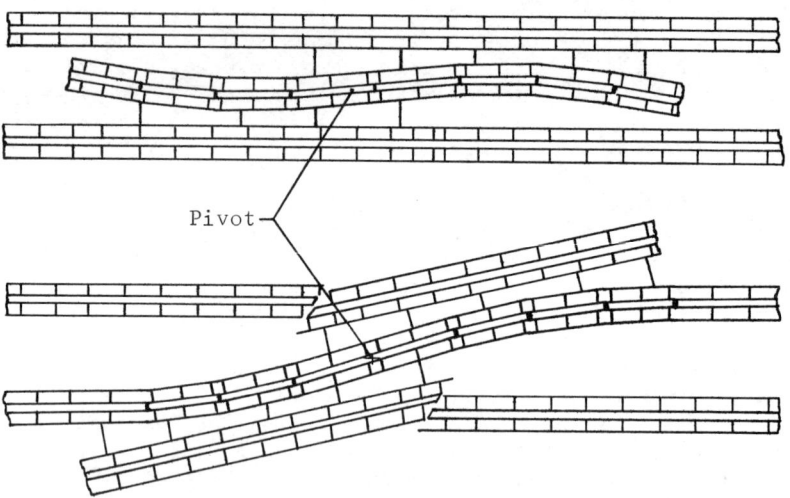

Fig 5. Pivotting switch

20. <u>Articulated Track Switch</u>. The most compact form of switch would be achieved if the track could be bent as in conventional railway points. The large angle through which the track must move and its rigidity prevent this approach being used. However, it is possible to construct the moving section of track from a series of short straight modules and to incorporate an articulated joint between each of the sections. Typically, nine modules would be required for a simple turn, with twice this number being needed for a two track crossover as in Fig. 4.

21. To preserve continuity a form of sliding joint would be necessary at the ends of the lift rails on each module and the electrical pick-up arrangement would need to cope with relatively wide gaps and, ideally, be able to bend as the modules move from the straight ahead to turn-out positions.

22. The modules are moved by either individual actuators or a single actuator acting through a complex linkage. Since the articulated modules have little vertical rigidity, each would require some form of sliding or rolling vertical support. The length would be the same as for a traverser, but the width does not exceed that required for the rest of the track.

23. The estimated cost of an articulated switch is less than that for a traverser, but the technical risk is much higher and the potential reliability is lower due to the large number of moving parts and the need for sliding joints on the lift rails.

24. <u>Pivotting track switch</u>. Fig. 7 shows a form of switch which has features from both the traverser and the articulated switch. In the full crossover configuration, three rigid sections of track linked together and rotate about their common centre in a similar manner to a railway turntable. Two of the tracks are straight and complete the route for travelling straight through. The third is S-shaped and forms the crossover. As the straight tracks swing away, the crossover moves into place. The ends of the moving and fixed tracks are shaped to keep gaps to a minimum.

Vertical loads are carried by the pivot point and by rollers at each end of the moving tracks which run on curved rails mounted on the track supports. Power to move the switch may be provided by linear actuators or by a rack and pinion arrangement.

This switch has fewer moving parts than the articulated switch and is more compact than the traverser. It has the lowest estimated cost of the three active options and represents a low technical development risk.

25. <u>Comparison between types of switch</u>. Whilst the passive switch itself offers advantages of low maintenance and high reliability, it necessitates the addition of a large amount of equipment on board the vehicles, increasing their cost and maintenance requirement and reducing reliability. Overall, it is less attractive than the active switch in almost all applications, the only exception being an unlikely system in which the number of switches is significantly greater than the number of vehicles.

Fig 6. The prototype switch in its alternative
positions.

26. All three active switches have particular advantages, with the pivotting switch offering the best compromise with regard to cost, reliability and technical risk. Accordingly, a prototype switch working on this principle has been designed and built, and added to the Maglev test track.

PROTOTYPE SWITCH
Track layout and switch geometry.

27. A full two-track crossover consists essentially of two simple turn-outs back to back, and the switching principle may be demonstrated by using just a single turn-out. The original Maglev test track at Derby was a single line at ground level and it was decided to add the switch at one end of this track. Due to restricted availability of space, a section of the existing track was demolished and a short length of curved track added to form the turn-out position for the switch. Two straight sections of new track running parallel with the existing line formed the straight-through position. Fig 6 shows the resulting layout, with the switch in both the straight and turn-out positions.

28. Design of the switch geometry concentrated on optimising the exact shape and position of the curved section of track and the location of the pivot point to minimise the overall size of the moving assembly. Computer aided design was used to speed up this iterative process and to check the loci of the moving ends of the tracks. Using a minimum curve radius of 20 metres, the total length of the switch is 12.5 metres.

Foundation

29. The foundation for the switch consists of a simple reinforced concrete raft on which all the switch components are mounted. In addition, it is extended to support the ends of the three fixed sections of track, thereby limiting relative movement and minimising the gaps that are necessary between the fixed and moving sections to allow for thermal expansion.

Mechanical Design

30. The original Maglev test track was constructed mainly from concrete to achieve a stiff track with a high resonant frequency and good damping. This was to prevent the possibility of interaction between the Maglev vehicle control system and the natural frequencies of the track. Later work, both experimental and theoretical, showed that lower frequencies would be acceptable. Therefore, steel construction has been used for the switch and for the additional lengths of fixed track. The same basic form of construction has been adopted for the Birmingham Airport link, so that a similar design of switch could be fitted retrospectively if required, or incorporated into any future Maglev installations.

31. Referring again to the cross section of the track shown in Fig. 1, the main bending loads are taken by the two universal beams. Their size is determined by the need to achieve the necessary stiffness over a 12.5 metre unsupported span, as intermediate supports on the switch are clearly undesirable. The chosen 406 x 178 x 67 section gives a

resonant frequency of 9 Hz, comfortably above the critical frequency. Compatibility with the vehicle and existing track dictates the size and relative positions of the support rails and the reaction rail for the linear motor. Bolted construction with shims is used extensively to allow tight tolerances to be maintained for the track assemblies without the need to select steel sections or to use jigs, and to allow final adjustments to be made easily. Limited use of welding minimises the risk of distortion being introduced.

32. Vertical loads at the ends of each of the two moving tracks are taken by rollers which bear upon curved tracks fixed directly to the concrete raft. The cross members linking the two moving tracks to form the switch assembly carry no vertical loads and are formed of rolled steel angle section. The pivot also carries no vertical loads and is mounted from the cross members.

33. The mating ends of the fixed and moving tracks are shaped to reduce the gaps to a minimum consistent with the allowance for thermal expansion. Shims between the switch assembly and the roller brackets allow precise adjustment of the height to ensure good vertical alignment between fixed and moving tracks.

34. Power to move the switch is provided by an electric linear actuator acting between a point approximately midway along the switch and a fixed bracket mounted on the concrete raft. A maximum force of 35 kN is available to enable the switch to be moved even with seized rollers. Transit time of the prototype is 30 seconds, although this could readily be reduced by fitting a more powerful actuator. Adjustable microswitches inside the actuator allow the extremes of travel to be set up within the specified tolerance of $\pm$ 5 mm.

35. Once the switch has reached either of its final positions, locking bolts extend from the ends of the fixed tracks into housings above the moving track lift rails. The wedges perform four functions.

a) Pull the tracks into final alignment from an error of 5 mm laterally and 3 mm vertically.
b) Provide vertical and lateral load transfer between the track ends.
c) Act as magnetic 'flux bridges' to minimise the effects of the physical gap in the lift rails on the magnetic suspension system.
d) To operate limit switches when fully extended to 'prove' the switch.

36. The wedges are moved by electric actuators mounted at the centre of the fixed track ends.

Electrical design

37. Electric actuators were chosen for operating the switch and the locking wedges because of the difficulty of providing any alternative power source and the ease of control. All are drive by three-phase induction motors supplied through contactors. Overcurrent protection is provided for the main

actuator whereas the locking actuators employ timer protection.

38. A simple relay control system is used to drive the actuators in the correct sequence and to monitor the limit switches. An automatic cycling controller has been added for experimental purposes to allow the switch to operate unattended to check its long-term performance.

39. The Maglev vehicle at Derby draws power from four conductor rails (3 phase and neutral) mounted in pairs beneath the track on either side of the centre-line. These continue through the switch, with a separate feed to the moving assembly. The gaps between the moving and fixed conductor rails are kept to the minimum permissible, and the ends are flared to ease the passage of the collector shoes. Two collectors per rail are used to minimise the risk of losing electrical continuity.

INSTALLATION

40. The various track sections were factory-assembled and delivered to the Derby site as complete units. Shims controlling the track gauge and the overall height of the track assemblies were already fitted.

41. Site installation was started by joining the short curved section of fixed track to the existing test track and fixing it down to the concrete raft with expanding anchor bolts, with packing being introduced to set the absolute track height. The designed level of the raft surface was such that nominally 50 mm of packing was required to allow variations in the actual surface level to be accommodated.

42. The moving track unit was lined up accurately using the previously laid track as a datum, and the exact positions of the pivot and the roller tracks determined. Finally, the two straight sections of track were installed either side of the switch, again using packing for height adjustment. The resulting gaps between the fixed sections of track and the concrete were grouted once satisfactory track levels had been achieved.

43. Commissioning consisted of fine adjustments to the tracks and adjusting the limit switches on the main actuator to set up the two end points of the switch travel. The limit switches in the locking wedge actuator were positioned so that the switch was only considered 'proved' if the wedges were in the fully extended position.

OPERATING EXPERIENCE

44. The switch has performed very satisfactorily during its period of commissioning and subsequent long-term tests, accumulating in excess of 10 000 operating cycles. No problems have been encountered with vehicle control system stability on the relatively flexible track sections, and the vehicle ride across the gaps in the track is good.

45. Some minor problems have arisen, in particular, the design of the locking wedges could be improved to give greater ability to overcome track misalignment and the limit switches inside the main actuator which determine the stopping positions have proved difficult to set accurately. An alternative means

of actuation, such as a rack and pinion arrangement coupled with limit switches operated by the track directly, may have significant advantages in future installations.

46. Although the benefits of bolted construction were appreciated during the fabrication and installation, partial replacement by welding could reduce the cost of the switch.

CONCLUDING REMARKS

47. A long-standing problem with Maglev systems has been solved by demonstrating a satisfactory switching device. The shuttle concept can now be extended to more complex systems, further increasing the applications of Maglev.

ACKNOWLEDGEMENT

48. The author wishes to thank the Director of Research and Engineering Development, BRB, for permission to publish this paper.

8 Rail corrugation—recent theories

R. A. P. CLARK, Research & Development Division, British Railways Board, Derby, UK

SYNOPSIS. A rolling load is usually expected to attenuate any surface irregularities, as is the case when rolling roads or lawns. Corrugation developing on smooth surfaces is a reversal of expected behaviour which requires a convincing explanation. This paper describes a number of theories which have been advanced in recent years to explain the widespread and apparently increasing phenomenon of railway corrugation. The essential features of the theories are presented and the extent to which they are capable of fitting the observations is discussed.

INTRODUCTION

1. Corrugations on railway rails are generally classified in two groups according to wavelength. Long wave corrugations have wavelength in the range 150 - 1200 mm with peak-peak amplitude up to 3 mm in severe cases. They are more likely to occur on curved track where they can be a derailment hazard but their chief significance lies in their ability to promote damage in structures, particularly bridges, if allowed to develop sufficiently. Long wave corrugation is not a serious problem in Great Britain, although it does occur; an example is shown in Fig. 1.

2. Short wave corrugations have wavelength in the range 40 - 100 mm; their peak to peak amplitudes rarely exceed 0.2 mm but this is large enough to generate considerable noise and track damage due to resonant effects, especially in sleepers (Ref. 1). Despite the fact that short wave corrugations occur where trains run at speeds as different as 5 m/s and 50 m/s the wavelengths are confined to the relatively small band 40 - 100 mm. Both straight and curved track is affected; an example on straight track is shown in Fig. 2 in which the pattern is fairly uniform along the rail. On high speed track the pattern is often confined to the second half of each sleeper bay as illustrated in Fig. 3; this feature seems to be confined to British corrugations and could be associated with the generally wider sleeper spacing used here.

3. Where corrugations occur it has seldom been possible to implicate any specific feature of rolling stock or track design or of the site itself. Many observations are of a

contradictory nature which suggests that a number of necessary conditions must be present in combination for corrugations to form. Despite the contradictions there does seem to be evidence that corrugation is more likely at sites where the atmosphere is damp and where the rails are supported on a particularly rigid foundation. It also seems that sites which experience a small variety of vehicles which pass with speeds in a narrow band are more likely to be affected. Short wave corrugation attacks different rail steels to different extents. In Britain rails made from Acid Bessemer steel are found to be more prone than other steels, (Ref. 2).

4. This paper lists a number of papers on the subject which have been published since 1950. It goes on to examine three theories in some detail and examines the areas in which they appear to be in agreement and disagreement with observations.

Above left: Fig. 1
Long wave corrugations on low rail of 200 m radius curve.

Above right: Fig. 2
Short wave corrugations on straight track.

Right: Fig. 3
Short wave corrugations on straight track, showing sleeper pitch modulation.
Direction of traffic: ⟶

REVIEW OF THEORIES IN THE LITERATURE

5. The extensive literature on corrugations dates from as long ago as 1895, and every author seems to have his own theory as to the cause of the phenomenon.

6. Inglis (Ref. 3, 1951) pointed out that the process of sliding a particle along a sinusoidal surface with friction was an unstable one in the sense that the irregularity would be accentuated. This is because the normal reaction between the particle and the surface, and therefore the wear, is greatest at the trough of the irregularity.

7. Turner (Ref. 4, 1954) thought corrugations were caused by the bouncing of wheels which led to dry and wet spots on the rail. Points where the wheels contacted became work hardened and the softer wet spots corroded faster than the harder dry spots forming the corrugation troughs. He suggested that uniformity of wheel diameter, braking action and rail steel composition were contributory factors.

8. Krabbendam (Ref. 5, 1958) and Werner (Ref. 6, 1975) considered ultrasonic vibrations of frequency 35,000 Hz or more in the railhead and wheels to have an influence on corrugation formation. The former also thought that internal stresses in the rail due to rolling and straightening had an effect.

9. Johnson and Gray (Ref. 7, 1975) investigated the hypothesis that corrugations can arise from plastic deformation of two surfaces in rolling contact. The wavelength of the corrugations they produced on a roller rig was determined by the rolling speed and the 'contact resonance' frequency which is the frequency the two rollers vibrated at on the Hertzian contact spring.

10. Clark and Foster (Ref. 8, 1983) investigated roll-slip vibrations as a mechanism by which short wave corrugations could be formed, and Mathieu type oscillations of wheelsets on a track stiffness variable at sleeper pitch as a mechanism for the formation of long wave corrugations.

11. Gasch et al. (Ref. 9, 1983) looked for a high frequency analogue of hunting which could explain corrugation formation. Their mathematical treatment concentrated on the possibility of exciting high frequency wheelset vibrations under certain contact conditions.

12. Engl and Meinke (Ref. 10, 1983) analysed the corrugations which formed on printing press rollers. By considering the system dynamics and wear behaviour they were able to recommend alterations which eradicated the problem.

13. The references above are only a small selection of those available. There is one factor, vibration, which is common to all of them. In the following sections, theories based on three different types of vibration are described.

LONG WAVE CORRUGATIONS RESULTING FROM TRACK STIFFNESS VARIATIONS.

14. Whenever a wheelset runs in flange contact with a rail there is the possibility of it vibrating laterally at a frequency determined by the unsprung mass and the lateral stiffness of the rail. Typically this frequency is in the range 6 Hz (for a locomotive with axle hung motors) to 20 Hz (for a trailer vehicle). If the wheelset has an angle of attack of less than about 10 mrads the mode will be heavily damped by wheel-rail creep forces. For angles of attack greater than this, full slip conditions exist and the wheel-rail forces are unable to provide damping; the mode will then be only lightly damped, by damping in the track structure.

15. One possible source of excitation of the vibration is rail joints. Another is the presence of a lubricator on the flange contacting rail. This can redistribute the friction forces acting on the wheelset and provide a sudden lateral force input. If the sleeper passing frequency, determined by the train speed, is close to the vibration frequency, the oscillation may persist due to the lateral stiffness variations in the rail at sleeper pitch (Fig. 4). There is one lateral oscillation per sleeper bay; persistent oscillations of 0.5, 1.5, 2.0 etc. oscillations per sleeper bay can also occur at the appropriate train speed.

16. The wheelset vibration can also occur on continuously supported rail but the lack of stiffness variation means that it is inevitably damped.

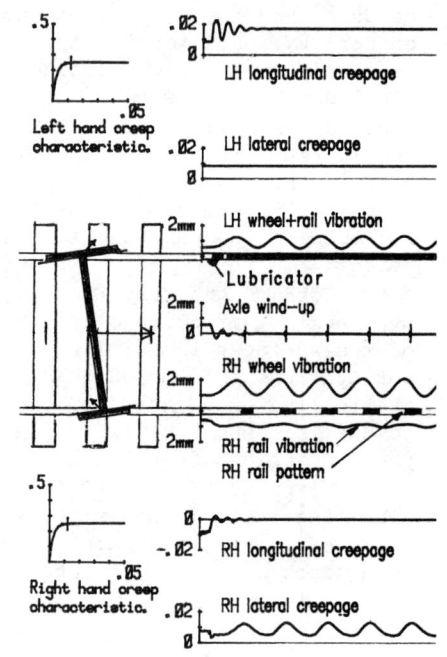

Fig. 4. Predicted response at 13 m/s

17. Factors encouraging this type of behaviour are:-
(a) The use of rigid non-steering bogies on sharp curves because this results in large angles of attack.
(b) Use of a rigid discrete rail support because this promotes large lateral rail stiffness variations and tends to reduce track damping.
(c) Large unsprung mass because this reduces the damping factor in the wheelset vibration for given levels of track damping.

18. Lateral vibrations of a wheelset have been produced under controlled test conditions which are described in Ref. 8.

ROLL-SLIP VIBRATION AS A CAUSE OF CORRUGATION

19. The idea that roll-slip vibrations are involved in the formation of some types of corrugation is quite an old one. It has been examined in some detail recently (Ref. 8). A necessary condition for this type of vibration to occur is that the friction force-slip characteristic at the wheel-rail contact has a falling characteristic at high creepage. The other necessary condition is that the creepage conditions are severe enough to give an operating point in the falling portion of the friction characteristic. If both of these conditions exist then considerable oscillation of the wheels and rails is expected; the mechanics of the vibrations are identical to those of violin string excitation.

20. The precise form of the vibrations depends on whether the applied creepage is in the lateral or longitudinal direction and on the train speed. When the creepage is lateral (as it is under angle of attack conditions) the vibrations are lateral oscillations in the non-flange contacting rail. An example is shown in Fig. 5. It has been found that the particular rail vibration mode depends on train speed in such a way that successively higher frequency rail modes are excited as speed increases; consequently the wavelength predicted by the roll-slip model is not very dependent on train speed.

21. Under high longitudinal creepage conditions set up, for example, by differing radii of the two wheels of a wheelset, the predicted oscillations involve the wheels vibrating out of phase on the torsional stiffness of the axle, as shown in Fig. 6.

22. Some experimental work has been undertaken to verify that roll-slip oscillations can occur in practice. This has

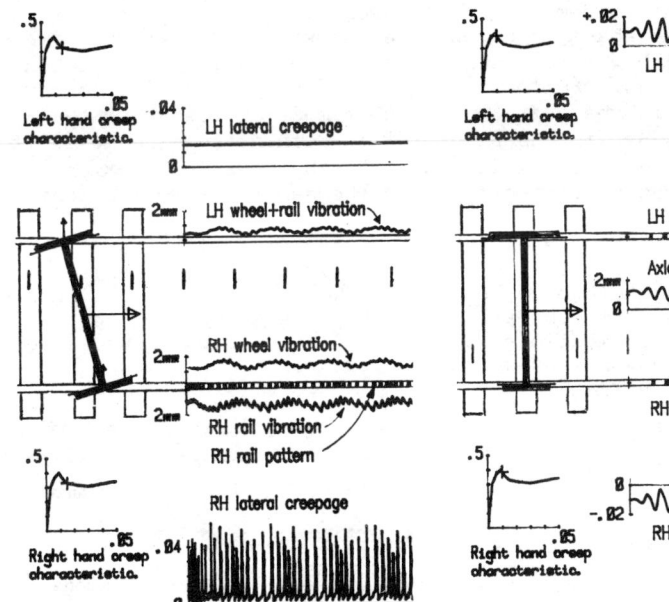

Fig. 5. Predicted response at 10 m/s under high angle of attack conditions.

Fig. 6. Predicted response at 15 m/s with large rolling radius difference.

involved running a Load
Measuring Wheelset under
creepage conditions above 1%
(i.e. a sliding velocity of
0.01 times the rolling velocity).
Considerable vibration did occur,
of a form consistent with the
predictions. In a particular
lateral creepage case the
vibrations resulted in a
pattern being left behind on
the rail, shown in Fig. 7.

Fig.7 Pattern produced in
high lateral creepage
conditions.

SURFACE ROUGHNESS DYNAMIC FILTER MECHANISM

23. Mechanisms for corrugation generation have been
proposed which make use of the fact that no rail surface is
ever completely smooth, but is randomly rough in the sense
that it contains small irregularities over a broad wavelength
range.

24. It was shown in (Ref. 7) that two rollers with randomly
rough surfaces sometimes develop corrugation when run at zero
creepage in a disc machine. The corrugations arose from the
periodic plastic deformation of the surfaces in association
with an elastic 'contact resonance' mode of vibration. Their
wavelength was determined by the rolling speed and the contact
resonance frequency, the latter being a function of the inertia
of the rollers and the stiffness of the Hertzian contact spring
joining them. Basically the resonance acted as a mechanical
filter on the random roughness, producing contact forces with
magnitude and phase to accentuate the wavelength in the
spectrum which was correct to excite the mode. It was found
possible to suppress the corrugation formation by increasing
the damping in the contact resonance mode, by reducing the
contact pressure to ensure operation within the elastic limit,
and by reducing the rolling speed so that the corrugation
wavelength was shortened to about the same size as the contact
patch.

25. Had the above experiment been repeated with a small
contact pressure but with some creepage it seems likely that
corrugations could have been formed with the same wavelength
but by a wear process rather than by plastic deformation.
Such an idea is mentioned in (Ref. 3) and has been successfully
employed in understanding the process of printing press bearer
corrugation (Ref. 10)

26. Railway wheelsets and
track are dynamically quite
complex. Nevertheless it is
possible to construct
mathematical models such as the
one shown in Fig. 8 and analysed
in detail in Ref. 1 and Ref. 11.
If the rail surface is assumed
to have a sinusoidal irregularity
of a particular wavelength, the
contact force will fluctuate

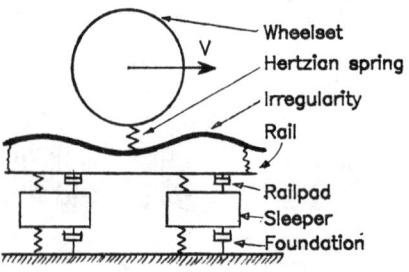

Fig.8 Vertical dynamic model.

harmonically once in each cycle of the irregularity. Under
non zero creepage conditions this produces a harmonically
varying wear deformation, while under high axle load conditions
it produces a harmonically varying plastic deformation. The
phase of the deformation in either case relative to the
irregularity is determined by the wheel and rail dynamics and
is important because it governs whether the irregularity grows
in amplitude or attenuates each time a wheelset passes. A
measure of the expected rate of growth of a particular
wavelength is given by the component of the fluctuating contact
force which is in phase with the irregularity. In carrying
out the calculation it is necessary to allow for the finite
size of the contact patch; this is because irregularities with
wavelength close to the length of the contact patch along the
rail will be attenuated irrespective of the system dynamics.
Results of such calculations for two train speeds are shown
in Fig. 9.

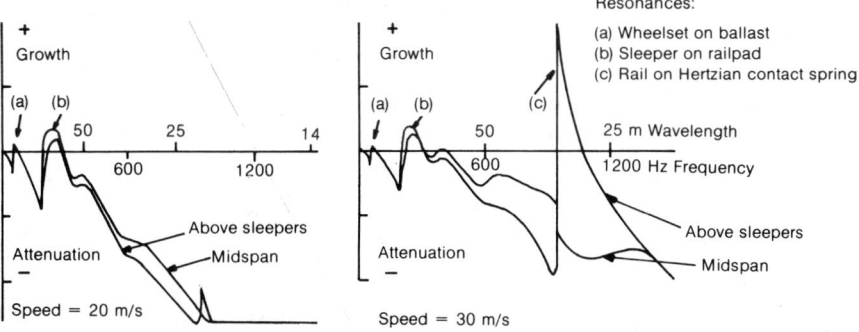

Fig. 9 Growth and attenuation of sinusoidal irregularities.

27. It can be seen that growth peaks occur at frequencies
associated with the various resonances. The heights of the
peaks, and therefore the maximum growth rates, are dependent
on the amount of damping in the track; smaller peaks are
predicted in well damped track. The high frequency growth
peak is absent at the lower train speed due to the attenuating
effect of the contact patch. This feature enables the model
to accentuate different frequencies according to train speed
with the result that predicted wavelengths are not particularly
speed dependent.

28. Time stepping solutions of the equations describing the
model may be carried out to predict the growth of a corrugation
pattern from surface roughness. Fig. 10 shows a typical
result.

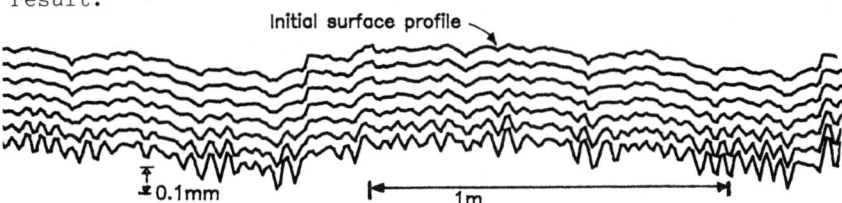

Fig. 10 Predicted corrugation growth from surface roughness.

CONCLUSIONS

29. All three corrugation theories described in some detail are able to predict vibrations which reproduce observed wear patterns. The roll-slip mechanism produces its most realistic prediction under high lateral creepage conditions and a difficulty arises in respect of whether sufficiently high lateral creepage occurs on straight track; some lateral creepage will undoubtedly be generated by wheelset misalignment and wheel diameter differences. The fact that only certain sites corrugate, and different steels exhibit different corrugation propensity could be accounted for by supposing the necessary friction characteristic and creepage conditions occur in some places and for some steels, but not in others. To avoid roll-slip vibrations, creepages should be kept moderate.

30. The dynamic filter mechanism is attractive because it needs lower creepage to operate than the roll-slip mechanism, and the creepage can be lateral or longitudinal. Differences between sites could be explained in terms of vertical damping differences but it is not clear at present how the differing behaviour of rail steels can be accounted for. In terms of this mechanism, corrugation avoidance is a matter of ensuring adequate damping in the wheel and rail vertical dynamics, and avoidance of high creepages and axle loads. Provision of a smooth initial surface on the rails also ought to be beneficial.

ACKNOWLEDGEMENT

31. The author wishes to thank the British Railways Board for permission to publish this paper.

REFERENCES

1. CLARK R.A., DEAN P.A., ELKINS J.A., NEWTON S.G. An investigation into the dynamic effects of railway vehicles running on corrugated rails. Jnl.of Mech.Eng.Sci. Vol.24 1982.
2. PEARCE T.G. The development of corrugations in rails of Acid Bessemer and Open Hearth steels. B.R. memo Aug.1976.
3. INGLIS C. Applied Mechanics for Engineers C.U.P. 1951.
4. TURNER T.H. Roaring rails. Rly Steel Top., Vol 2 No. 3 1954.
5. KRABBENDAM G. Is rail corrugation due to internal stresses? Bull. Int. Rly. Congr. Ass., March 1958.
6. WERNER K. Corrugation and pitting of rolling surfaces - are they contingent on ultrasonics? Wear, Vol 32, 1975.
7. JOHNSON K. L., GRAY G.G., Development of corrugations on surfaces in rolling contact. Proc. I. Mech. E. Vol 189 1975.
8. CLARK R.A., FOSTER P. Mechanical aspects of rail corrugation formation. Paper presented at Tech. Univ. Berlin, 1983.
9. GASCH R., GROS THEBING A, KNOTHE K., VALDIVIA A. Linear self excited vibration as initiating mechanism of corrugation. Paper presented at Tech. Univ. Berlin, June 1983.
10. ENGL A., MEINKE P., Corrugation on bearers as effects of the short time dynamics investigated in the long time wear process. Paper presented at Tech. Univ. Berlin, June 1983.
11. GRASSIE S.L., GREGORY R.W., HARRISON D., JOHNSON K.L. The dynamic response of railway track to high frequency vertical excitation. Jnl. of Mech. Eng. Sci. Vol 24 No. 2 1982.
12. CLARK R.A., FOSTER P. An investigation into the effect of vertical dynamics on corrugation formation. BR memo Feb 1984.

9   Modern in-track rail head rectification

J. COOPER, Speno International Geneva, Switzerland

SYNOPSIS. The critical position of the rail is emphasized.
The wide range of rail head surface defects that can be
treated by in-track rectification are described with reference
to UIC definitions. An explanation is given of the latest-
generation rectification units and their controlled
application in the longitudinal and transverse planes. Special
mention is made of rectification in switches and crossings. A
model is outlined for evaluating and optimizing rectification
strategies. Areas meriting further development work are
pinpointed.

CONTEXT

1. Seen in cross-section, the vehicle-track system appears
as two extremely large bodies – one moving, one fixed –
interacting at an extremely small contact surface. The change
in magnitude of the pressures through the fixed support
reveals how the rail bears the brunt of the forces generated
by the moving mass.

upper surface	relative bearing pressure
rail	4,000
ballast	40
subgrade	1

2. The absorption of load takes place over a depth of some
50cm most of which is composed of ballast that has been
vibrated into place. Service vibration generated by unevenness
on the rail head surface will tend to deconsolidate this vital
element of the track structure. Thus, in the longitudinal
plane the rail head surface must be maintained to within tight
limits.

upper surface	relative accuracy
rail	1,000
ballast	10
subgrade	1

3. In the transverse plane demands on the rail head surface are just as high. Effective conicity — defined by the marriage of rail profile and wheel profile — determines vehicle behaviour and in extreme cases derailment protection. In this critical environment small deviations can have far-reaching effects. Surface defects are expressed in tenths of a millimeter in an environment of thousands of kilogrammes and hundreds of metres per second. This perspective emphasizes the potential negative cascade effect of rail head surface defects and the sensitive nature of rail head rectification.

DEFECTS

4. The UIC Catalogue of Rail Defects (ref.1) describes most rail head surface conditions that can be treated by modern rectification equipment. These defects include deviations from longitudinal and transverse profiles and general surface conditions, such as :

Defect	UIC No.	Remarks
transverse cracking	211	from local dynamic effects
short-pitch corrugation	2201	3 - 8cm long
long-pitch corrugation	2202	8 - 30cm long
long waves	—	up to 200cm long
lateral wear	2203	bevelling of gauge face
abnormal vertical wear	2204	with formation of lip
surface defects	221	flaking, etc
shelling	2221	on running surface
shelling	2222	on gauge corner
crushing	223	with formation of lip
local batter	224	with formation of lip
isolated wheel burn	2251	possibly after resurfacing
repeated wheel burns	2252	with superficial cracking
bruising	301	for slight cases
wheelflat damage	—	for slight cases
surface pitting	—	for slight cases
irregular welds	—	possibly after lifting

RECTIFICATION

5. The Speno rectification process consists of grinding with the flat end of a rotating annular wheel. Several wheels are applied simultaneously to each rail. The wheels can be linked longitudinally to establish a reference base some 250cm long. The transverse rail head profile is reproduced in the form of a multi-faced polygon by the action of grinding wheels operating at various angles. An example of values for this geometrical approximation on a BS113A rail (before initial traffic "rounds off" the facets) :

	gauge corner	transition	running surface
design profile radius	12.7mm	79.4mm	304.8mm
facet width	1.5mm	3.5mm	7mm
number of facets	11	4	3
angle between facets	6.5degrees	4.3degrees	2.7degrees
max.dev. profile-polygon	0.02mm	0.02mm	0.02mm

6. The aims of the rectification process are to :
- re-establish the longitudinal profile of the rail head surface
- to remove metal that has reached or is near its fatigue limit
- re-form a suitable transverse rail head profile.

7. The ideal longitudinal rail head surface profile is a smooth surface in a theoretical straight line. This result can be virtually achieved for defects up to some 180cm long. Above this length the rectification process is no longer absolute and gives diminishing returns of metal removal. However, deviations from the longitudinal profile at these wave lengths also become less critical and a technical-economic balance indicates an acceptable residual long wave depth of some 0.3mm.

8. The target transverse profile is less easily defined. The question needs further thought by railway engineers. Restoring the original design profile may involve extra work for an effect that is quickly lost. An average wear profile may be too generalized (particularly as rectification is a process destined for zones with specific, non-average problems).

9. Considerations to be borne in mind in determining a suitable profile include :
- local profile deformation history
- the avoidance of concentrated contact stresses
- the respect of effective conicity limits
- the possible influence of a given profile on the formation of rail head corrugation in the longitudinal plane
- the immediate cost of achieving the profile.

10. On site the elements for a - constantly changing - rational choice are not to hand. While awaiting clarification the practitioners have opted to grind a transverse profile that :
- has no obvious lip
- allows optimum removal of longitudinal defects
- leaves a wheel-rail contact band some 25mm wide centered over the rail web (except for deliberate asymmetric grinding to reduce lateral rail wear).

EQUIPMENT

11. Fig. 1 shows how the Speno grinding units - each consisting of a grinding wheel, an axial transmission and an electric drive motor - are mounted in longitudinal pairs about a transverse central pivot point (1). The cradle containing the pair of units is further mounted on longitudinal pivots (2). The adjacent angle between the two units can be set mechanically from 0 to 20 degrees. The whole cradle can be tilted to give grinding angles up to 75 degrees on either side of the vertical. Special units with shaped grinding wheels are used for attaining the gauge corner behind the check rail in switches and crossings. Two controls are available in the machine cabin : instantaneous blocking and unblocking of pivot

point (1) depending on the length of the defect being treated, and continuous tilting of the cradle about pivots (2) in order to adjust the transverse rail head profile being formed. Switch and crossing rectifiers also have a control for holding units clear when traversing the running surface gap at the crossing nose.

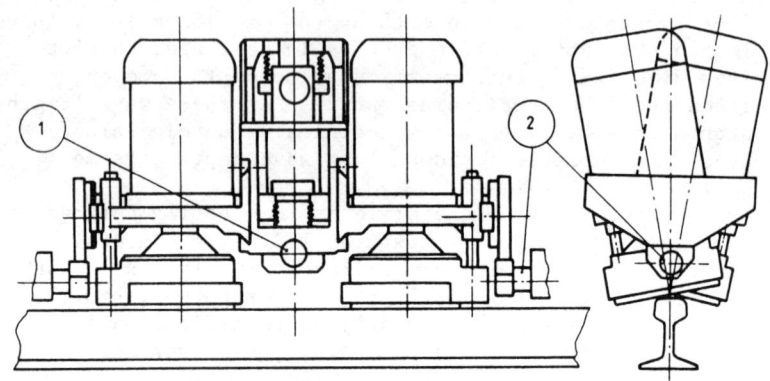

Fig. 1   Grinding group

12. The pairs of units are mounted – in sets of four or eight units – in retractable trolleys slung under carrying vehicles. The total number of units on a machine can vary from 16 to over 100 depending on the application in mind – from delicate work in switches and crossings to high-production plain line performance. All machines are self-propelled.

13. Use of these powerful rectifiers implies an accurate rail head measuring system. For the longitudinal plane the machines incorporate devices for recording the depth of surface irregularity (with linked filters for distinguishing between short and long waves). The information is retained in the form of analogue traces on a multi-channel line recorder in the cabin.

14. Continuous measurement of the transverse rail head profile is more difficult. Fig. 2 illustrates the Speno multi-point contact shoe now being introduced for this purpose. The signals from the displacement transducers on the shoe are relayed to a VDU screen in the cabin.

CONTROL

15. Speno International is engaged in a development program designed to automatize the rectification process. Most elements are now in place. The next stage will be the linking of the components into an effective cybernetic loop.

16. Already control in the longitudinal plane is quasi-automatic. The application pressures of the grinding wheels are maintained at preset values by a system that senses the current intensities of the individual electric motors and

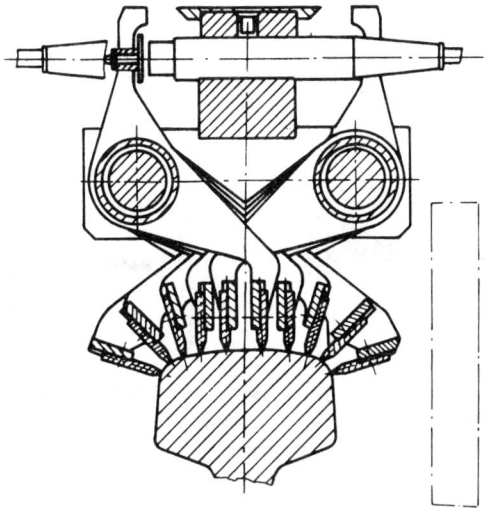

Fig. 2   Transverse profile measurement

increases or decreases the pressure in the application
cylinders accordingly. The effect of the successive grinding
passes can be evaluated from the traces on the line recorder
and a timely decision taken for breaking off the work (subject
ideally to a visual check in track). As the traces distinguish
between short and long waves they provide guidance information
on the - at present human - decision to block or unblock the
pairs of units.

17. When working on plain line it may be necessary to raise
locally individual units to clear certain types of track
apparatus. The manoeuvre is facilitated by an obstacle monitor
that, when actuated - at present by an operator - memorizes
the nature and the position of the obstacle and raises and
lowers successively the units concerned. It is planned to
extend this device to cover the memorization of complete
switch and crossings layouts to ensure sure passage of
successive grinding passes.

18. For the moment the operator remains a vital link in the
control of the rectification process in the transverse plane.
Information on the current rail head transverse profiles is
relayed from the measuring shoe on each rail into the cabin
where it is compared with a preset, electronically-memorized
target profile. Fig. 3 shows the two forms of VDU profile
comparison for guiding the operator in his choice of grinding
angles. To simplify this task preset multiple angle patterns
can be called up by numeric code. It is intended that
recognition of profile deviation and setting up of
corresponding grinding angles will be taken over by a
computer.

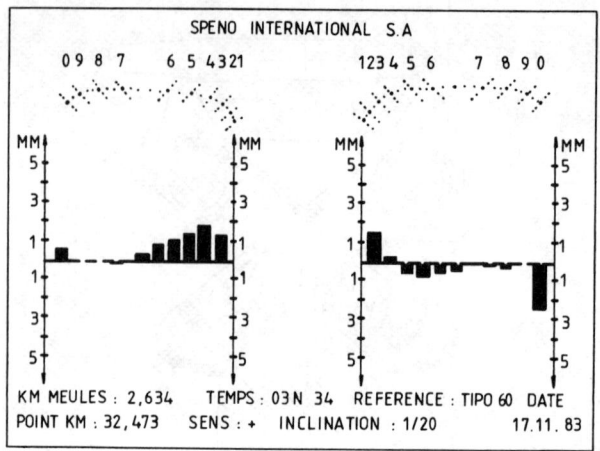

Fig. 3.   VDU Display

MANAGEMENT

19. Correct management of rectification equipment implies
knowing when and where to grind. A fundamental choice when
setting up grinding programs is the steering factor
determining the rectification intervention. The factor may be
cyclical in nature — years of rail service, or tonnes of total
train mass — or a systems dynamic approach may be preferred on
the basis of measured rail head defects. The latter method
requires the regular reconnaissance of long stretches of
track. Speno has in service a self-propelled recording vehicle
for this purpose. Known as the SM775, this machine currently
records rail head defects in the longitudinal plane (including
switches and crossings). A longer-term aim is to fit
additional equipment on this machine to gather simultaneously
information on the rail head transverse profile.

20. Basically, deciding a rectification strategy amounts to
setting a threshold value on the chosen steering factor. The
decision translates into an instruction to rectify a given
zone, say every year, or every million tonnes, or when the
rail defects need, say, five grinding passes for elimination
(ref. 2). Threshold values cannot be determined simply. Rail
head defects have a widespread effect on the railway system
and many aspects must be taken into account.

21. Joint research on rectification strategies is being
undertaken by the Deutsche Bundesbahn (German Federal
Railway), the Technical University of Hanover and Speno
International (ref. 3). Entitled "Determination of the optimum
technical-economic point in time for rail rectification as a
component of preventive track maintenance", the project
incorporates three computer models for exploring :
- the relationship between track parameters and rail head
defect growth

- on-site operations research
- the cost-benefit balance of steering factor thresholds.
  22. On the costs side the last model covers :
- programming
- payment of rectification
- associated railway expenditure
- traffic disturbance.

23. The benefits are based upon the avoidance of extra costs due to rail head defects. The notions retained are :
Material : rail amortizement
Track maintenance : tamping, fastenings, ties, ballast cleaning
Social costs : noise, vibration
Vehicle maintenance : wheel turning, general overhaul
Exploitation : passenger loss, speed restrictions.

24. The object of the model is to establish the relationship between intervention threshold levels and resulting total system net costs (expressed as monetary units per metre of track per year). It is hoped that this relationship will exhibit a reliable minimum value representing the optimum intervention threshold. (The model does not include preventive grinding of new rail).

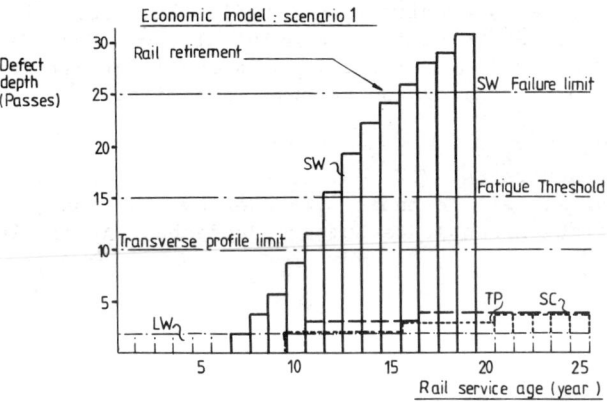

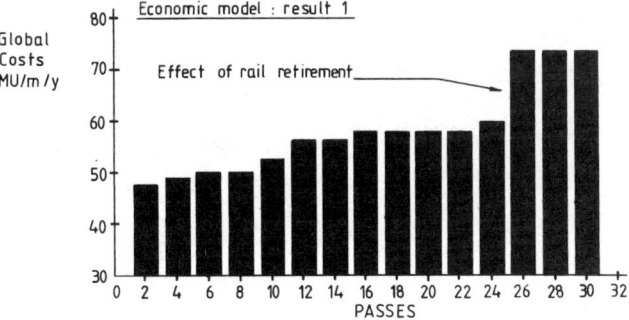

Fig. 4 Test scenario and result

25. Work on this model (which is intended for extension to other trackwork operations) has highlighted the need for railways to invest in adequate cost records and financial procedures. Difficulty has been experienced in describing adequately some of the benefit notions. Opinion polls have been used for several minor elements. However early sensitivity analysis indicates that the relationship is dominated by a small number of better-known major elements and that tentative conclusions should be available.

26. Fig. 4 shows a prototype test scenario and result that illustrate the important effect of rail amortizement costs.

CONCLUSIONS

27. Until recent years rail rectification was not accorded the attention that it merits. Practised principally for correcting surface defects in the longitudinal plane, rail grinding risked stagnating as a local operation of a stopgap nature. However increased awareness of the importance of maintaining the transverse rail head profile and a growing realisation of the economic ramifications of rail rectification led railways to press for more flexibility and more control of this vital operation. The foundations are now being laid for an era of fully controlled, almost-automatic rail rectification. Progress has been made in the development of new equipment. Greater understanding has been achieved of the critical nature of the operation. The next decade should see in-track rail head rectification realising its full potential of technical and economic benefits.

REFERENCES
1. UIC Brochure No. 712, ed. 1979
2. O'Rourke, Rail Technology, Frederick and Round, 1982
3. Fendrich, Eisenbahntechnische Rundschau, April 1984

Discussion on Papers 6–9

MR F. I. MAU, Civil Design Engineer, BHP Engineering, Sydney

Mr North, what are the cost per metre and the carrying capacity of the Birmingham Maglev system?

MR B. H. NORTH

The cost of track depends on the requirements of the particular installation and is heavily influenced by site conditions and operating or safety requirements (such as the need for a central walkway). The track at Birmingham was designed at an early stage in the vehicle's development before vehicle tests or trials were undertaken. The design loading given was conservative but now, with the results of later work available, we are preparing designs for a much cheaper structure for future applications.

An application is currently being considered in which the peak capacity is about 10 000 passengers per hour on a single track.

MR A. WINTER, Senior Development Assistant, London Transport

Mr Lilley, why was the vehicle suspended from an underhung magnet rather than from a simpler mounting on the roof which would make point and crossing work less complicated?

MR M. J. LILLEY

A Maglev vehicle must have some form of additional structure to support it and to bring it safely to rest in the event of failure of the magnetic levitation system. If the suspension were from an overhead track, the emergency support system would probably consist of a set of pads or rollers mounted on struts above the vehicle which bear on the upper surface of the track in the event of a de-levitation. The need for these struts to pass through the track at points and

crossings would introduce similar problems to those
encountered with the present arrangement.

An emergency system consisting of pads or rollers beneath
the vehicle would be expensive as an additional track or
roadway would be required at ground level. Also it would be
very difficult to design, as the tolerance in height between
the main track and the emergency track would need to be very
tightly constrained since the 'flying height' is typically
only 15 mm.

Even if it did prove somewhat simpler to engineer a
switch for an overhead track, the additional cost and
complication of the overhead structure would far outweigh any
savings on the switch.

DR -ING J. EISENMANN, Professor of Civil Engineering,
Technische Universität München

Mr Lilley, the Birmingham Airport Maglev has no secondary
suspension between the car body and the magnets. How good is
the riding quality at the maximum speed of 75 km/h?

MR M. J. LILLEY

The normal ride height of the Maglev vehicle is 15 mm, and
with the quality of track specified for the Birmingham Airport
link (which approximates to good quality secondary railway
track), the root-mean-square (r.m.s.) suspension movement is
typically ± 4 mm at the maximum speed of 13 m/s.

The ride quality has recently been measured and values of
1% g r.m.s. vertically and 3% g r.m.s. laterally were recorded
at the top speed. An improvement in the lateral value is
expected following modifications to the control system.

DR -ING J. EISENMANN, Professor of Civil Engineering,
Technische Universität München

SHORT PITCH CORRUGATIONS

SUMMARY
The theory is based on the residual stresses caused by
manufacturing and changed under traffic-added load stresses,
which lead to a flow process with a viscoelastic effect in the
zones near the edge.

It is well known that under operating loads the rail is
hardened to a depth of 6-8 mm from the contact point of the
wheel by

(a) the bending compressive stress and additive stresses
caused by disturbed bending (Fig. 1)

(b) the compressive stresses especially directly under the
wheel as a result of local concentrated loads as well as in
the longitudinal and transverse direction of the rail head and
in the vertical direction (Fig. 1).

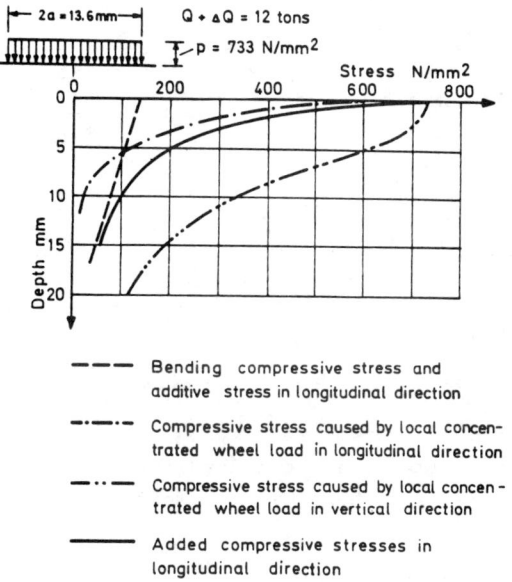

Fig. 1.   Stresses in the rail head under the wheel

In addition 2 mm under the 20 ton axle the compressive stress
in the longitudinal direction of the rail head averages about
500-600 N/mm^2 (Fig. 1); the maximum value (P = 99.7%) is 50%
above that.   For a 30 ton axle the longitudinal stress
increases to 640-740 N/mm^2 (mean value).

   With repeated wheel loads in the zones near the edge as
well as in the longitudinal and transverse directions of the
rail head these stresses, which exceed the yield point and far
exceed the elastic limit of the steel, lead to compressive
residual stresses.   Consequently, with the new straightened
rails the tensile residual stresses (Fig. 2) in the edge zones
in the longitudinal direction are gradually reduced and the
compressive residual stresses are activated.   The area of the
change in stress extends from the rail head to a depth of
about 30 mm, and in the centre of the rail head there is a
zone where there is no reversal of tension (increase in
tensile residual stress) (Fig. 2).

   The stress reversal which appears in the longitudinal
direction of the rail head in the area of the upper 6-8 mm
results in an added compressive residual stress which
increases during operation.   On top of the original residual
stress when the rails are straightened, and a very marked

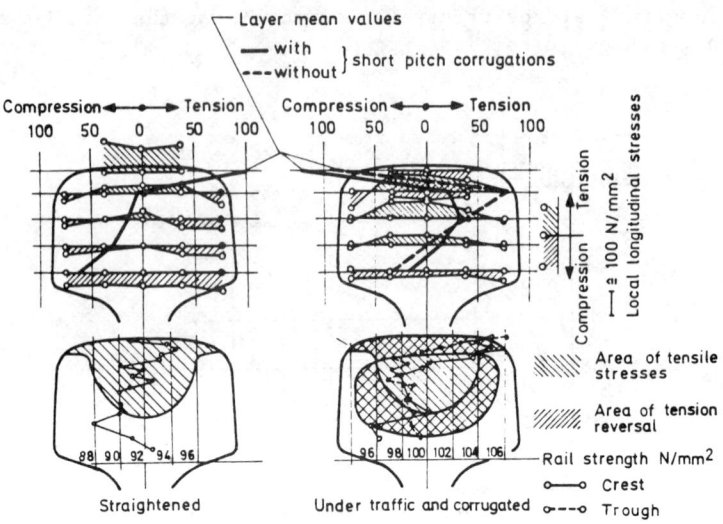

Fig. 2. Strength (given by hardness RH$_c$) and longitudinal residual stresses of a straightened rail and an S54 corrugated rail under traffic with 900 N/mm^2 ultimate tensile stress

plastic deformation of the zone near the edge, the so-called Bauschinger effect is evident with stress reversal (Fig. 3). This produces a slight flow process for relatively small stresses with opposite sign. Also, this effect is evident for a repeated loading in the same direction.

The longitudinal compressive stresses from the traffic load are overlaid by the gradual build-up of compressive residual stresses. At 2 mm depth these are approximately 600–800 N/mm^2 after operation with a 20 ton axle load and 850 mm wheel diameter; for a 30 ton axle load the added longitudinal compressive stresses are 740–940 N/mm^2. The added compressive stress increases by 60–100 N/mm^2 in the summer when the temperature is high, in both a continuously welded track and a track with joints. This produces a locally limited flow process in the zones near the edge which progresses with repeated loading. Through the effect of the longitudinal pressure on both sides of the flow zone and the effective compressive residual stress vertically underneath, the flow area bulges, corresponding to the crest of the corrugation. Of significance is the viscoelastic effect present during the flow, which manifests itself in a phase shift between the force which has been introduced (effect of the wheel) and the deformation, i.e. the reverse deformation releasing the crest occurs after a time lag after the wheel has passed. At the same time, the adjacent pressure-relieved area decreases owing to negative transverse elongation, corresponding to the

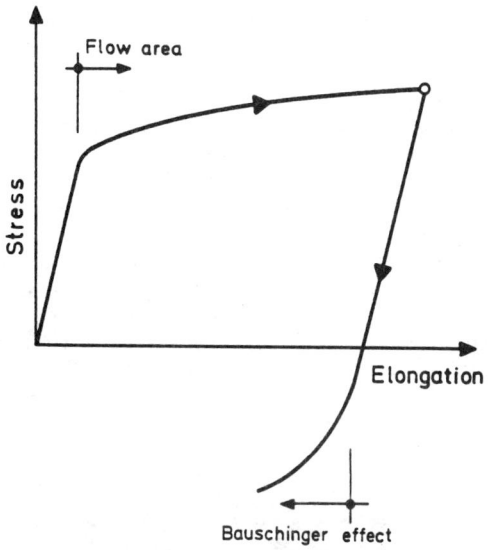

Fig. 3. Bauschinger effect illustrated in a stress-strain diagram

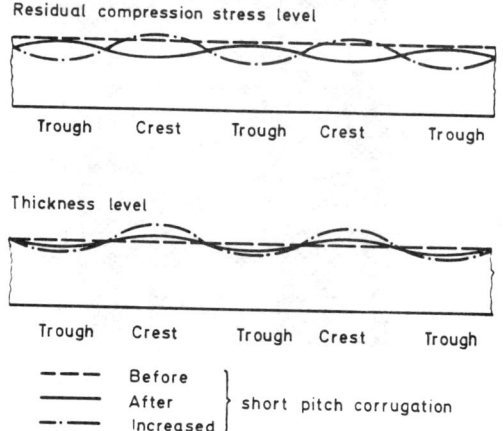

Fig. 4. Residual stress and thickness of the zone near the contact area

corrugation trough (Fig. 4). The compressive residual stress created as a result of the plastic deformation decreases slightly.

After the short pitch corrugation has been formed the wheel touches only the crest of the corrugation, to give the black and white grid typical of corrugations, in which the black spots correspond to the trough no longer touched by the wheel. Under progressive load action the very fine martensite layer on the surface breaks and leads to the black spots visible on the corrugation crest in Fig. 5. Under this load increased by the dynamic process the compressive residual stresses rapidly increase again in the area of the crest, along with a progression of the flow process as well as hardening, and under the longitudinal pressure of the neighbouring zones the crest arches again at the same time that the trough falls. The difference in height between the crest and the trough rises to 0.15 mm, 0.20 mm and up to 0.40 mm.

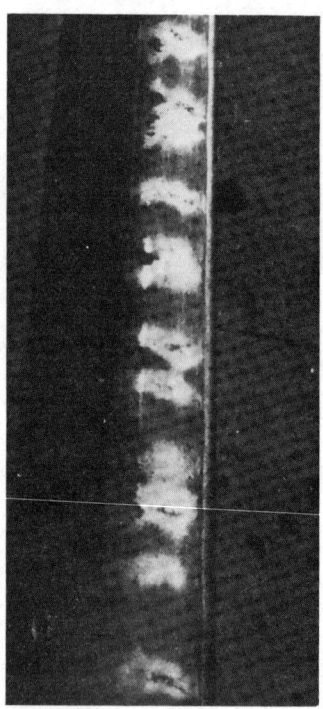

Fig. 5. Short pitch corrugated rail (right-hand side: traffic edge)

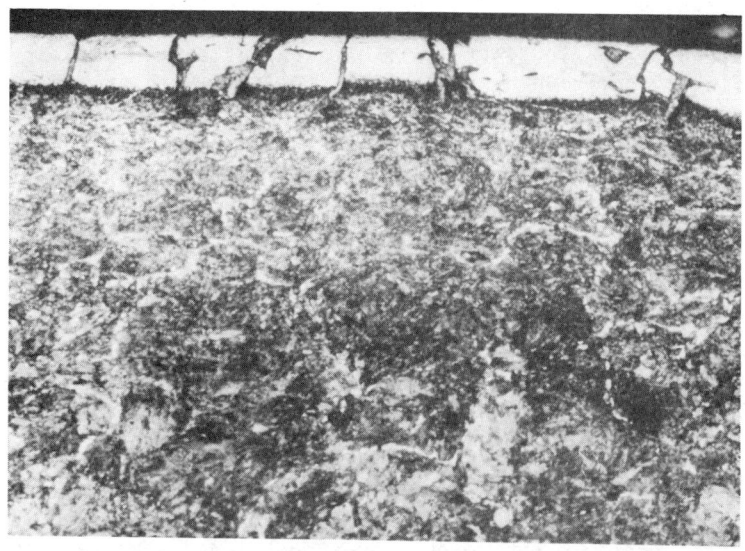

Fig. 6. Arching of the corrugated crest: opened joints in the thin zone of martensite

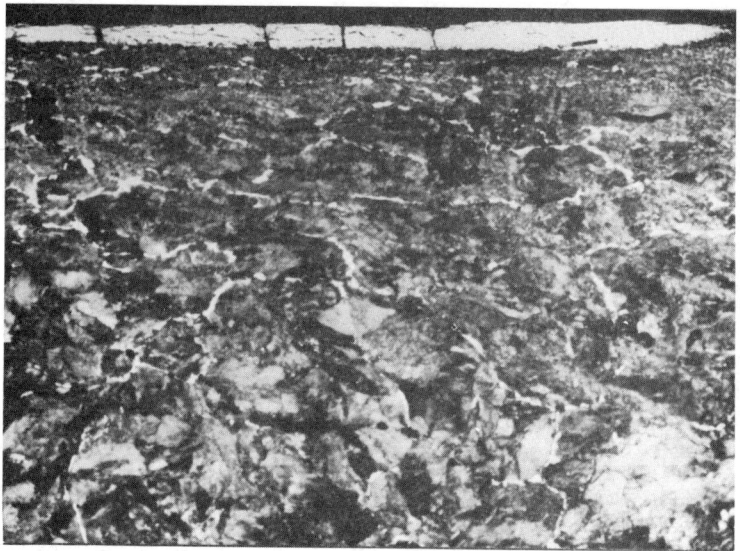

Fig. 7. Corrugated trough: closed joints in the zone of martensite

As shown, arching is connected with plastic deformation and sagging of the bordering area with an elastic deformation when the pressure release is simultaneous (reduction in the compressive residual stress). Arching of the surface of the rails is confirmed by structural studies, which show a widening of the fracture gaps in the very fine martensite layer on the surface, in the area of the crest, whereas the fracture gaps are closed in the trough (Figs. 6 and 7).

The length of the flow zone, corresponding to the crest, can be explained by the tension field generated under the wheel. Hence, for normal axle loads and wheel diameters the length of the contact area between the wheel and the rail is approximately 15 mm. The stress field present in 2-4 mm depth, corresponding to the flow zone, is about twice as long, so that for the arch the length is about 25 mm and the ridges are 50 mm apart. The distance between ridges in the operating track is 30-70 mm.

What can we do to prevent short pitch corrugation? The answer based on this theory is 'nothing', but if we grind the new rails after laying and welding, the corrugation starts later because the smoother surface leads to a decrease in the contact pressure (no concentrated peaks) and therefore the plastic flow near the rail edge is reduced. The next recommendation is that, if we grind the rails shortly after the corrugation starts, it is possible to stop the development of the very marked flow process and the build-up of a hard martensite layer 0.2-0.3 mm thick.

BIBLIOGRAPHY

1. Krabbendam, G. Internationales Schrifttumsverzeichnis uber Riffel- und Wellenbildung auf Eisenbahn- und Strassenbahnschienen. Forschungsdienst der Niederlandischen Eisenbahnen, Utrecht, 1961.

2. Eisenmann, J. Theoretische Betrachtungen uber die Beanspruchung des Schienenkopfes am Lastangriffspunkt. Eisenbahntechnische Rundschau 14 (1965), No. 1, pp. 25-34.

3. Oberweiler, H.G. Ein Beitrag zur theoretischen Untersuchung und experimentellen Prufung von Eisenbahnschienen. Mitteilungen des Prufamtes fur Bau von Landverkehrswegen der Technischen Universitat Munchen, No. 17, 1973.

4. Meier, H. Eigenspannungen in Eisenbahnschienen. Organ fur Fortschritte im Eisenbahnwesen 8 (1936), No. 15, pp. 320-329.

5. Yasojima, Y. and Machii, K. Residual stresses in rail. Permanent Way No. 26, Tokyo.

6. Dogneton, P. Contribution a l'etude des phenomenes de fatigue de rail; le phenomene de la 'tache ovale'. Dissertation, University of Paris, 1971 (see also Schienen der Welt, 1972, Dez., pp. 877-881).

7. Muller, H. Unpublished work.

8. Eisenmann, J. Formation of short pitch corrugations in rails. Heavy haul Railways Conference, Perth, 1978.
9. Leykauf, G. Riffelbildung bei Schienen - Versuche. Seminar Fahrweg, Augsburg 1979. Deutsche Eisenbahn Consulting.

MR R. A. CLARK

The subject of rail corrugation is from a theoretical point of view quite involved and so it is always risky to comment on other people's theories without being familiar with all the details. It is possible that corrugations arise from a number of distinct physical effects and several theories may eventually be required to fit all the observations. Any theory that leads to the conclusion that corrugation is inevitable is at variance with the observation that, although corrugation is common, it is by no means universal and some stretches of line are never affected.

MR J. M. CRUDEN, Canadian Transport Commission, Hull, Quebec

With regard to paragraph 3 of Mr Clark's paper, while with East African Railways and Harbours (1953-65) I relaid a large amount of main line. One length in Kenya had had a history of 'roaring rail' but had been ground and was being fitted with tapered shims at rail joints when I took it over. The track was 80 BS rail on steel sleepers which had pressed-up lugs to take key rail fasteners. The ballast was no better than adequate and there had been a problem with alkaline corrosion of sleepers. Fatigue crack propagation was· evident at many of the sleeper lugs and the resultant looseness of keys entailed annual creep correction work, as no rail anchors had been installed. Track drainage was good and in excellent repair. The locality was semi-arid, dry except during monsoon rains. The subgrade was silty, sandy murrum.

The track was rebuilt with BS95R rail on punched steel sleepers taking A, B, K clip fasteners (allowing gauge widening increments, also metre to 3 ft 6 in gauge conversion) secured by high tensile tee-head bolts, with about 12 in new ballast under the sleepers and a ballast shoulder of at least 12 in. All joint and fastener bolts were lubricated with grease-graphite compound when installed and annually thereafter. All curves were stringlined, corrected and beaconed at 31 ft centres for future maintenance, with necessary bank widening preceding ballasting. Rail joints were squared, except staggered in curves. There was no problem with rail corrugation in the 5 years after reconstruction. All trains were limited to 45 m.p.h. maximum but reconstruction was in anticipation of an increase to 60 m.p.h. maximum.

Do rails from acid Bessemer steel have a markedly different natural frequency from others?

MR R. A. CLARK

Natural frequencies of railway lines are determined by mechanical properties such as bending stiffness, density and sleeper spacing and support. It is not therefore expected, nor is it observed, that rails of acid Bessemer steel would have different natural frequencies from rails of other steel, all other things being equal.

We believe that the tendency for acid Bessemer rails to be affected by corrugation to a greater extent than rails of other steel is the result either of a different balance between the wear and work-hardening properties or of the tendency to form a hard surface layer such as white phase. Such factors are likely to affect the tendency of vibrations induced by many successive wheels to fall into step and to cause the build-up of a discernible pattern.

DR S. L. GRASSIE, Research and Development Engineer, Pandrol Ltd, London

The principal problem to be confronted in providing a mechanism which explains why short wavelength corrugations develop is that these corrugations have a wavelength which is typically in the range 40-80 mm and which varies little with train speed. Whereas constant frequency phenomena are common in mechanical engineering, constant wavelength phenomena are rare. It may be, as proposed by Mr Clark, that the complicated dynamic system of the wheel and rail responds in a higher frequency mode the higher the speed of the train in such a way that the distance travelled in one oscillation of the system varies within quite narrow limits. The fundamental mechanics of this selection are not understood at present, nor has it been demonstrated that the track has sufficiently poorly damped resonances at the required frequencies.

An alternative hypothesis which has been under investigation at Cambridge University Engineering Department in collaboration with British Rail involves interaction of creep forces at the wheel-rail contact with the longitudinal dynamics of the system. The creep, or traction, force is a function of creep $e = v/V$ where v is the relative velocity of wheel and rail at the contact patch and V is the forward velocity. For small relative velocities (typically much less than 1% creep) the traction $f = Ce$ where C is the creep coefficient. Thus the contact can be regarded as a dashpot of strength C/V. If this dashpot is in series with a spring of stiffness s, this simple dynamic system responds with a characteristic length $l = C/s$.

This simple characteristic has directed our thoughts.

Unfortunately the two stiffnesses which most directly present themselves have proved unrewarding in providing us with a plausible corrugation mechanism. The contact stiffness which arises from local elastic deformation is too high and gives too short a characteristic response length. The longitudinal track stiffness is of the correct order, but the track longitudinally is also a very strong dashpot and is massive: consequently the system is unresponsive.

The fundamental idea is attractive because of its simplicity and because the mechanism would always be latent. It may be too simple.

MR D. SHARPE, Chief Civil Engineer, Mass Transit Railway Corporation, Hong Kong

In your paper, Mr Clark, you declare that 'despite the contradictions' there are four reasons for increased corrugations: dampness; rigid foundations; small variety of vehicles; speeds in a narrow band. Apart from dampness are not the other three truisms?

MR R. A. CLARK

In the large amount of published material on corrugations there are many contradictions concerning factors which influence their formation. It therefore seems worthwhile to list the areas of agreement. Four factors are often mentioned: dampness, rigid foundations, small variety of vehicle speeds and speeds within a narrow band. From an engineering viewpoint the last two must make matters worse, whereas perhaps the first two would have more subtle influences. However, from a theoretical point of view the importance of the last two factors is not immediately obvious since the wavelength of observed corrugations is largely independent of whether the predominant train speed is high or low, and most of the vibrations which are predicted to take place occur in the rails, not in the vehicles.

Theoretical studies lead to the conclusion that although particular factors influence corrugation growth there is no single parameter which controls the process; rather it is a question of the existence of a number of necessary conditions. Perhaps this prevents an easy engineering intuition of the importance of particular features and this is one reason why the understanding and prevention of corrugation has proved to be time consuming and difficult.

DR C. ESVELD, Head of Railway Technology and Quality Control, Nederlandse Spoorwegen, Utrecht

Mr Cooper, from a number of recent papers dealing with heavy haul railways, asymmetrical grinding seems to be beneficial for increasing rail life in curves. This has also been confirmed by Mr Fahey.

Does the Speno company have experience with this new technique on mixed traffic railways? What do you expect the benefit will be in this case?

MR J. COOPER

We are not aware of any 'scientific' trials of asymmetric grinding under mixed traffic conditions in Europe. In North America a set of carefully followed trials was undertaken by Speno Rail Services on the Canadian Pacific network. A detailed report of these trials was issued by Mr S. T. Lamson of the Canadian Institute of Guided Ground Transport.

We believe the technique of asymmetric grinding is worthy of research in mixed traffic conditions and the Speno Company would be pleased to co-operate in field trials.

While awaiting convincing evidence, our impression is that asymmetric grinding could bring some relief in lateral rail wear on mixed traffic routes in curves where flange contact can be influenced by acting on the rolling radius element (possibly in the 500-1000 m curve radius range).

10

Track maintenance planning based on quality information produced by the NS track recording system BMS

Dr. C. ESVELD, Head of Rail Technology & Quality Control, Permanent Way Department, Netherlands Railways (NS), Utrecht, The Netherlands

SYNOPSIS. In 1983 the new track recording system BMS, developed jointly by NS and Delft University of Technology, has been taken into service. BMS records track geometry up to wave lengths of 70 m, as well as quasi static components. The signals are processed on-line by an IBM computer producing, inter alia, quality indices per 200 m section.
The NS network has been split up into 600 maintenance sections. Besides 3 quality classes are distinguished. The quality rating system Q-NORM produces a maintenance priority list, based on BMS data.

SELECTING THE MEASURING SYSTEM

1. The new NS system came into being as the result of collaboration between NS and Delft University of Technology. In 1978 a preliminary study was made of all the systems available at that time, followed by a more detailed study of the British Rail track recording system, which in 1979 led to the decision to adopt a non-contact measuring system based on an inertial measuring principle [1].
Adopting such a non-contact measuring system according to the "stabilised platform" principle has two big advantages when compared with conventional systems. In the first place, there is no wear, as there are no mechanical transducers. This results in an enormous reduction in maintenance costs, as 90% of the maintenance costs of a conventional system are due to the mechanical transducing system. The second advantage concerns the much larger measuring range - BMS measures wavelengths up to 70 m - and the undistorted recording of the track geometry.

THE MEASURING PRINCIPLES

2. BMS has been designed in such a way that the measuring speed may vary between 40 and 180 km/h. The lower limit is related to the reduction in the acceleration signals and thus to the measuring accuracy at reduced speed. Whenever the speed drops below 40 km/h the computer gives a signal to this effect. The system measures the following magnitudes: cant, level, alignment and gauge in the 0.5 - 25 m and 0.5 - 70 m wave bands, while the quasi-static part, represented by the 70 - ∞ m wave band is also determined for cant, curvature and gauge.

The track twist is calculated from the difference in cant on a
basis of 6 m and 2.75 m.

3.   How this method works broadly is to be explained by means
of the level. The measured vehicle body acceleration is integrated twice with respect to time in order to obtain the absolute
car body displacement in space. Naturally, the measuring range
always has a finite range. Hence the long wave lengths are cut
off electronically at 70 m. The purpose built electronic control system tunes the parameters defined on the basis of time
continuously with speed, so that the measuring characteristic
on the basis of spatial frequency and wave length is not changed. As the accelerometer is mounted rigidly on the floor of
the car body the acceleration is always recorded perpendicular-
ly to the floor, whilst in actual fact it is concerned with the
vertical acceleration. This is derived from the measured signal by making an electronic correction as a function of the
car body rotation due to the cant. The displacement between
rail and body is subtracted from the absolute body displacement thus determined, giving the vertical level of the track.
To measure track cant the variation of the car body rotation
per unit of time, measured with a so-called rate gyro is used.
In the lateral direction the variations in displacement are
carried out in two stages, i.e. between rail and bogie frame
using line scan cameras and between bogie frame and car body
using linear displacement transducers.

As a result of the body rotation there is a rotation in the
accelerometer. In order to prevent gravity causing a virtual
lateral acceleration, the measuring system electronically eli-
minates the gravity component.

Figure 1 gives a diagram of the system used to determine the lateral displacement between rail and bogie frame. With the help
of a mirror galvanometer a laser beam is projected 14 mm underneath the rail head. A line scan camera with an array of 256
diodes determines the position of the projected point from the
reflected light and thus determines the relative position between rail and bogie frame.

BMS MEASURING PRINCIPLE GAUGE AND ALIGNMENT

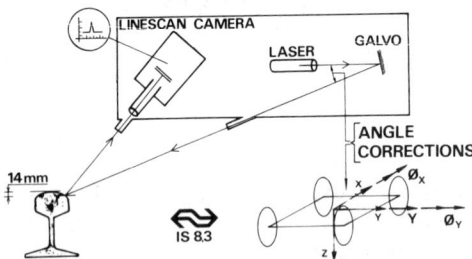

Fig. 1

To provide the mirror galvanometer with the exact information so
that the angle may be adjusted continuously, 4 linear displacement transducers are mounted between the bogie frame and the
axle.

Altogether 16 magnitudes are measured by one rate gyro, 2 acce-
lerometers, 10 linear displacement transducers, 2 horizontal
line scan cameras and 1 tachometer.

ON LINE PROCESSING OF THE MEASURED DATA

4.   The instruments which have been mounted in the BMS and
also the relationship between the different components have
been reproduced in figure 2. The measuring system supplies analog signals which are processed on-line by a Series 1 type IBM-
computer. This processing means that for each 200 m section
standard deviations and from these quality indices are determined, whilst the IBM machine also records whether any of the
signals exceeds certain levels. This computer also carries out
the task of computing the track twist from the cant recording.
Since the figures from the previous run are available on floppy disc, also the difference with respect to this run, i.e. the
deterioration or improvement, is printed.

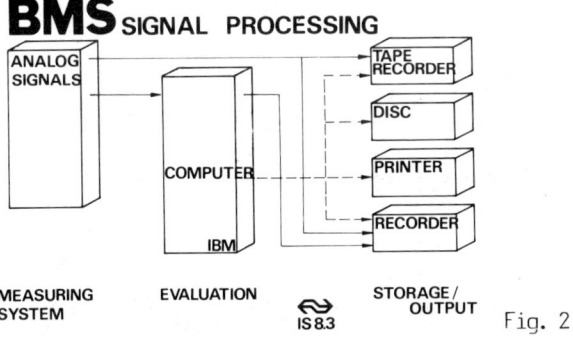

Fig. 2

5.   The IBM Series 1 computer has both a hard and a floppy
disc unit on which the programs and computed data are stored
respectively. The analog data is stored on tape via a 14 track
FM tape recorder (1" tape width) of the type Thorn EMI SE 7000.
The results of the on-line data processing carried out during
the recording run are displayed on two printers. One of these
gives the quality indices and the other prints the local defects. The analog signals are displayed graphically on a "Gould"
recorder, amongst others.
The examination for local defects is limited to cant, level,
alignment and twist in the wave bands 0.5 to 25 m. For errors
in cant, level and alignment an exceedence level is used of 6
σ 80 (σ 80 is the standard deviation per 0-section to be explained afterwards, with a minimum of 8 mm, whereas the value
for track twist amounts to 10 mm. Whenever the system detects
that one of these level-1 standards has been exceeded, then
this is printed by location and peak value. Whenever the peak
value also exceeds level 2, which is set at 20 mm for all signals, the track maintenance department should immediately take
remedial action.
Early 1984 the system will be enhanced to measure corrugation
and bad welds for optimising the use of a rail grinding train
and the STRAIT system.

QUALITY RATING

6. Quality indices and maintenance sections

The NS network has been split up into about 600 maintenance
sections (O-sections), which can be possessed for track maintenance. Besides 3 quality classes (Q-classes) are distinguished.
Q-class 1 concerns mainline tracks with maximum speeds $v \geqslant$ 130
km/h. Tracks with 100 km/h $\leqslant v <$ 130 km/h and mainly consisting of CWR tracks have assigned Q-class 2, whilst all jointed
tracks and CWR tracks with $v <$ 100 km/h are considered as Q-
class 3.

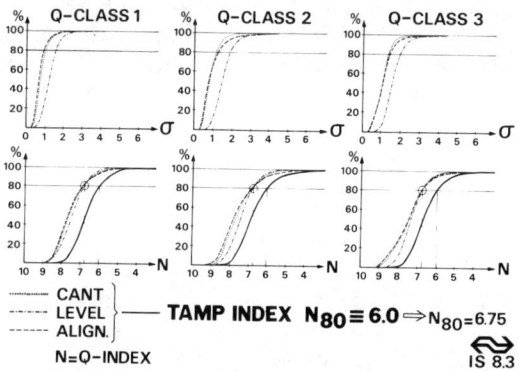

Q-NORM GEOMETRY NS-NETWORK SPRING 1983 **BMS**

Fig. 3

On the basis of the BMS compaign conducted in spring 1983, the
cumulative distributions of the standard deviations per 200 m
section for cant, level and alignment in the 0.5 - 25 m wave
band, as well as for alignment in the 25 - 70 m wave band, have
been determined in each of the 3 Q-classes. As can be seen from
figure 3, the shapes of the different distributions show little
similarity. To achieve that the curves possibly well coincide,
the standard deviations have been transformed to quality indices, as indicated in Figure 4.

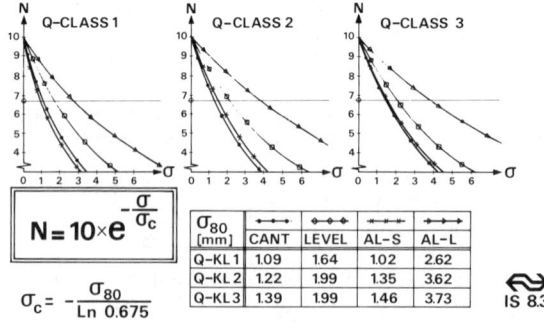

Q-NORM RELATIONSHIP BETWEEN N AND σ PER 200m **BMS**

$$N = 10 \times e^{-\frac{\sigma}{\sigma_c}}$$

$$\sigma_c = -\frac{\sigma_{80}}{Ln\ 0.675}$$

σ_{80} [mm]	CANT	LEVEL	AL-S	AL-L
Q-KL1	1.09	1.64	1.02	2.62
Q-KL2	1.22	1.99	1.35	3.62
Q-KL3	1.39	1.99	1.46	3.73

Fig. 4

The quality indices range from 10 (best) to 0 (worst); the
maintenance intervention level corresponds to an index 6. The
values σ 80, used in figure 4, are the 80% -points of the
preveously mentioned cumulative distributions for the network
per Q-class. All these 80%-points by definition coincide now,
as demonstrated by figure 3, and thus all quality indices for
each signal and each Q-class have the same meaning, especially ambient to the 80%-points.

7. Producing a track maintenance priority list.

From the point of view of riding comfort and track forces, the
quality of the various geometry components for each O-section
should not only be as good as possible, but also as uniform
as possible. Since, in principle, all Q-indices have the same
significance, the lowest index of the three indices pertaining
to cant, level and alignment is retained as tamping index for
each 200 m section. By way of indication it roughly applies that
if the tamping index is lower than 6, maintenance may be necessary, However, the decision will not be based on one single 200
m section, because for this purpose O-sections were created.
The step necessary to determine from the 200 m sections a quality index for each O-section proceeds as follows. The tamping
index for each 200 m section will not be constant throughout
the length of the O-section. So as to also allow for this variation, not the mean value but the value at 80% is adopted,
which means that 80% of the 200 m sections in a O-section will
have a higher and 20% a lower index. The Q-index per O-section
in the wave band 0.5 - 25 m thus established, furnishes the final assessment as far as automatic tamping/lining is concerned.
The Q-NORM-system [2] selects the 600 O-sections in the sequence of this tamping index; the O-section with the lowest index
is found at the top of the list and the O-section with the highest index at the bottom.

Q-NORM   MAINTENANCE ADVICE

PRIORITY LIST MECHANIC MAINTENANCE MAINTENANCE SECTIONS TOTAL NETWORK BMS RUN [SPRING 83]							IS 8.3
O-SECTION from to	L km	Q cl	Q-INDEX 0-25 meter			TAMP	ALIGN- LONG
			CANT	LEVEL	ALIGN		
211.A.03 Ltn-Ed	6.8	2	6.0	6.3	4.6	4.6	3.5
026.B.02 Odzg-Odz	6.2	1	5.9	5.8	5.4	4.9	6.0
107.A.02 Mda-Ztmo	5.4	1	5.0	6.5	6.8	5.0	7.2
115.A.01 Sdm-Vdgo	4.4	1	5.5	5.5	5.2	5.0	4.2
132.B.03 Rtd-Rtng	3.4	1	5.6	5.7	5.4	5.0	4.8
115.B.01 Vdg-Sdm	4.6	1	5.9	5.8	5.2	5.1	3.7
132.B.02 Rtdg-Nwk	7.4	1	5.1	5.8	5.7	5.1	5.5
226.A.02 Wad-Apn	8.0	2	5.4	5.9	5.9	5.1	6.1
200.A.04 Wfm-Rd	5.0	2	5.4	6.3	5.5	5.1	4.1
103.A.02 Apn-Ledn	6.2	1	5.3	6.5	5.3	5.1	5.3

N 80

Fig. 5

Figure 5 gives the first part of the priority list calculated
early in 1983 from the BMS measuring run.
Figure 6 shows the list of priorities in a condensed form. These
distributions, based on the BMS-run of spring 1983, have been
used to determine the N 80 value assigned to the 80% points of
the earlier mentioned national distributions displayed in figure 3. This has been done by applying an iterative procedure,
such that in the priority list pertaining to the O-sections,

20% of the length of the network considered has, by definition, a tamping index less than the value 6. The amount of 20%
has been chosen because NS yearly tamp approximately 20% of the
network.
Apart from the tamping index dictating the sequence, the system calculates, in an analogous way, the 80% value for cant,
level and alignment in the 0.5 - 25 m wave band and also the
80% value for the alignment in the 25 - 70 m wave band. This
latter index gives an indication of the need for design lining.

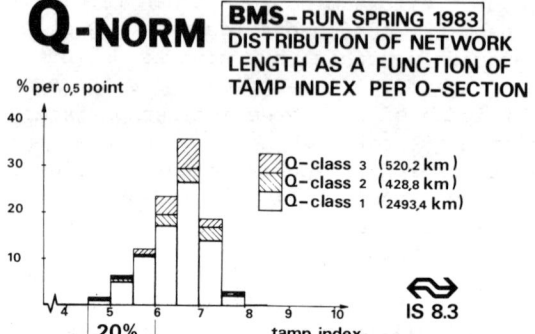

Fig. 6

The Q-indices for each 200 m section are calculated directly
by the IBM computer of BMS during the recording car run. The
priority list according to figure 5 can only be elaborated after completing the recording car campaign. It is intended to
introduce the extrapolated figures into the Q-NORM-system. i.e.
the figures in which the deterioration over 6 months has been
introduced.
    8.   The introduction of O-sections implies, in fact, a choice
in favour of maintaining long consecutive line sections. However, also in the case of renewals, a treatment per O-section
should be the objective. Only when these principles will have
been adopted in all the processes, will this approach of treatment per section bring optimal results from the point of view
of quality and effeciency.

EXPECTATIONS FOR THE FUTURE
    9.   About 10 years ago the Permanent Way Department of NS
started on digital analysis of geometry signals, recorded by
the track recording car. In the past, this was done after the
completion of the measuring campaign, but for the new BMS-system
the analysis is made directly during the measuring process.
Apart from information concerning short waves to 25 m length,
BMS also furnishes data on long waves to 70 m length and quasi-
static signals for cant, curvature and track gauge, being particularly useful in the assessment of curves.
The great problem in the analysis of geometry measurements lies
in the combination of the various results to an overall figure.
The method described above, though furnishing a solution,
lacks physical substantiation. Moreover, the assessment of the
track quality cannot be considered without allowing for the rolling
stock (see Appendix). After all, vehicle response magnitudes, such

as car body accelerations and wheel/rail forces decide whether passengers comfort and stresses in the superstructure respectively remain within permissible limits. The direct measurement of vehicle response for assessing track quality has considerable drawbacks. The results of the measurements are strongly speed depend and only apply to one specific type of rolling stock. Moreover, there is again no information concerning the track geometry. Therefore, track geometry recordings, independent of speed, will always continue to constitute the basis. Considered from this point, vehicle respons should be determined from the geometry by calculation.

10. The gist of the problem thus lies in finding an answer to the question how the relationships between track geometry components on the one hand and vehicle response on the other can be determined in a practical way. To this end NS has chosen the MISO methode (Multiple Input Single Output), which has also been adopted by the ORE C 152 Specialists Committee since 1979. For further details regarding this system, reference is made to [3]. According to the MISO-approach, schematically illustrated in figure 7, the geometry components cant, level, alignment and gauge are considered to be vehicle excitations (4 inputs) and the car body acceleration as response (1 output).

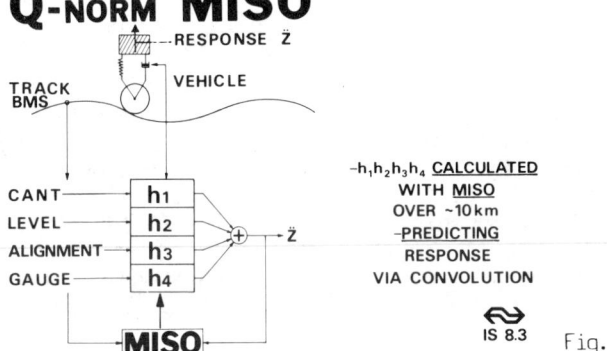

Fig. 7

The geometry components and also the vehicle response are measured and recorded on magnetic tape, from which the MISO-system calculates the transfer functions between geometry and response. Precise details on the procedure followed in this method are given in the following example: the track geometry has been measured over a distance of 10 km with the aid of BMS and, simultaneously, the vertical car body acceleration of the NS measuring car has been recorded. Using the MISO-method, the transfer functions between the 4 geometry components and the car body acceleration of the NS measuring car have been computed from these data. Subsequently, the car body acceleration has been calculated, or in other words predicted, from the measured geometry, using the previously estimated transfer functions, and compared with the real, i.e. the measured car body acceleration.

Figure 8 shows the measured and calculated vertical car body acceleration as a function of the distance covered. As shown by the graphs, the two values display very good concord. Recent tests have shown that the results, also laterally, though slightly less good, are all but acceptable.

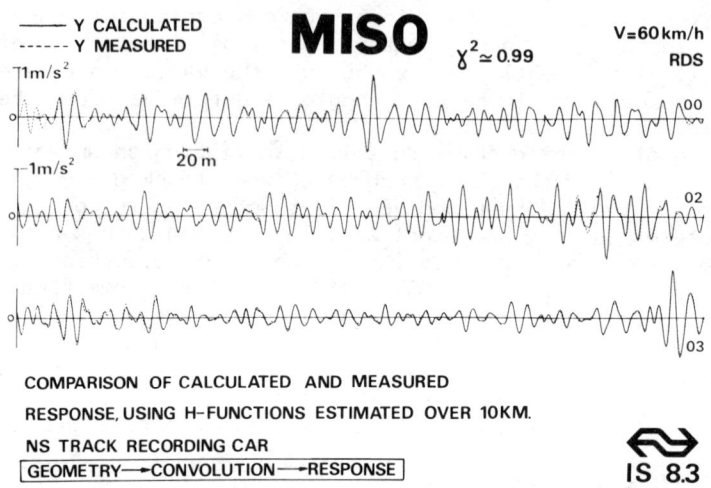

COMPARISON OF CALCULATED AND MEASURED
RESPONSE, USING H-FUNCTIONS ESTIMATED OVER 10KM.
NS TRACK RECORDING CAR
GEOMETRY—►CONVOLUTION—►RESPONSE

IS 8.3

Fig. 8

In fact, the problem of experimentally determining the relationships between track geometry and vehicle response has been solved following the completion of the MISO-system. In fact now only the actual performance of measuring the response on various types of rolling stock at different speeds, together with the track geometry, so as to enable the transfer functions to be determined by MISO, is still pending.
These tranfer functions will have to be incorporated into the present BMS-system. Here the use of a number of micro-processors connected in series is considered; they should enable the various vehicle responses to be calculated on line via the convolution principle. As may be expected, this realisation will take a few more years.

References

[1] Esveld, C.: "NS adopts contact free measurement of track
                geometry", Railway Gazette, November 1983.

[2] Esveld, C.: "Q-NORM, Quality rating of track geometry",
                Rail International, June 1984.

[3] Esveld, C.: "MISO: Application of random signal analysis
                to the interaction between vehicle and track"
                Rail International, August/September 1984.

Appendix

Some considerations on track quality.

When looking at track irregularities different wave bands can be considered. Faults in the wave band 0 - 3 m are mainly due to rail shape and welds, whereas the longer waves are originated from ballast and sub soil. Therefore it will not be possible to represent the track quality by one simple figure. Besides it is not the geometry which is decisive but the vehicle response, as only forces and accelerations can be tested against physical standards.

			TRACK MAINTENANCE				
			GRINDING		TAMPING		
			STRAIT		AUTO	DESIGN	
WAVE BAND [M]				0–3	3–25	25–70	70–∞
FREQUENCY BAND [Hz]		1000–250	140–0				
GEOMETRY	CANT						
	LEVEL						
	ALIGN.						
RESPONSE	Q–FORCE	?	?	///?///	/////////	/////////	?
	A–BOX ACC.						
	BODY ACC. (COMFORT)				/////////	/////////	?
TRACK QUALITY		CORRU-GATION	BAD WELDS	ROLLING DEFECTS	SHORT + LOC. DEF.	LONG	CURVES
			RAIL		BALLAST		
ASSESSMENT CRITERIA: ISO,STRESSES							

‾‾‾ DIRECTLY MEASURED BY TRACK RECORDING CAR **(BMS)**

///// CALCULATED FROM MEASURED GEOMETRY **(MISO)**

? STILL TO BE STUDIED

IS 8.3

The above figure tries to represent the previously described ideas, which have been or shall be realized in the NS track recording system BMS. The irregularities are split up according to the maintenance processes available for correcting the track geometry. The long wave irregularities are measured via the non-contact system described in the main text. Conventional track recording systems mostly restrict to the wave band 3 - 25 m. BMS additionally records long wave information and quasi-static information. A further essential step, which is mostly omitted, consists of calculating vehicle response data from the recorded geometry. As the long waves are primarily related to car body vibrations they are well representing car body accelerations and thus passenger comfort. The transfer functions necessary to determine these car body accelerations from track geometry can be established via the MISO method. NS presently test a new implementation for on-line calculation of vehicle response through recursive filters. This method has the great advantage of substantially reducing the computational time compared to ordinary convolution. A further point of study concerns the realisation of the filters in hardware.

Traditionally the track quality is practically exclusively assessed on the basis of information provided by conventional track recording systems, i.e by looking of the long waves. Rail induced irregularities, like corrugation, rolling defects and bad welds are in this way underestimated, although they are however responsible for the high frequent forces and impact loads, which from a major aspect in the track deterioration phenomenon. Because of the great importance of the short wave irregularities NS started the development of extending BMS with an axle box acceleration measuring system for monitoring bad welds and rail surface deviations. In addition also the vertical geometry in the wave band 0 - 3 m is monitored for each rail.

11   What is expected from track?

G. H. COPE, FICE, MCIT, BSc(Eng), ACGI, Permanent Way Engineer, British
Rail Headquarters, London, UK; R. J. GOSTLING, MA, MIMechE, Head of
Track Research Unit, Railway Technical Centre, Derby, UK

SYNOPSIS. Although the track is one of the fundamental
elements upon which a railway undertaking is founded, each of
the various protagonists in the railway business sees the track
in a different way and expects different things from it. This
paper discusses the nature of these various expectations about
track and the way it performs, and the way they interact with
one another.

INTRODUCTION
Who expects what?

1. Railway track has several functions. Notably it acts:

   - as support for traffic loading

   - as guidance system

   - as carrier of signalling messages

   - as fuel conveyor

2. For this reason a wide range of parties both within and
without the railway organisation itself, take an interest in
the track, and have expectations about how it must perform in
relation to their own particular sphere of activity. A list
of such bodies would include:

   - The Board

   - Sector Directors

   - Functional Directors

   - Department of Transport

   - Customers

   - Neighbours

3. It is impossible within a short paper to deal with the
expectations of all who have interest in the track. This is
the more so due to the diverse nature of the interests at
stake. These interests include:

   - First cost

- Maintenance/Renewal Cost
- Speed
- Load Carrying capacity
- Operational convenience
- Safety
- Freedom from environmental intrusion
- Signalling
- current carrying

4. The topics which on examination prove to be of general concern to the majority and/or the most significant classes of interested parties are those to do with the general flexibility of infrastructure management in relation to the railway business. Putting it crudely and cynically, the ideal railway track would be one that cost nothing to lay, never required any maintenance and carried any desired load at any speed. In the real world however decisions have to be taken about the factors of cost, maintenance, speed and load which are mutually interdependent and it is from the economic and technical conflict between these factors that this paper derives its theme.

What Kind of Track?

5. No one paper could cover all the track expectations for railways throughout the world. The problems would for example be different if one considered construction of a new railway from what they would be if one were considering the adaptation of an old one. In fact we propose to limit ourselves to B.R. This means we start with an existing layout and a large stock of rails, sleepers etc, that we want to get the best out of. This means that an expectation which required wholesale re-railing (for example) is likely to be doomed from the start.

Expectations

6. The expectations about track on B.R. today are personified so to speak by the recently devised management framework of Sector Directors - Inter-City; Freight; Provincial Services; London and South East (LSE); Parcels. Each looks at a different bit of the railway (with some overlapping) and each view is different. A broad generalisation of outlook might be:

- Inter-City and LSE are concerned with speed
- Freight and Parcels with carrying capacity
- Provincial Services and LSE with low cost
- all are concerned with cost and availability.

We will therefore be concerned with expectations on these four subjects as they apply to the existing railway.

MAXIMISING SPEED CAPABILITY

General

7. B.R. has been in competition ever since its inception, either from alternative railways in the past, or from road and air at present. Speed is therefore an issue in the passenger carrying railway, as along with convenience and cost it is a key factor in how or even whether a journey is made.

8. It becomes a problem, as distinct from an issue, because the running of fast long-distance passenger trains over the same tracks as short distance passenger trains and freight trains at lower speeds causes capacity difficulties. Operation at higher speeds also raises potential technical problems. The former aspect (which can be thought of alongside ideas of achieving maintenance without disrupting traffic) i.e. the desire to maximise availability of the permanent way, will be considered later. Only the technical aspects of high speeds on curved and straight track, will be considered in this section.

Curving Behaviour

9. Historically the difficulties inherent in running at differing speeds round curves were resolved by a compromise between what were understood to be the ill-effects of excessive cant deficiency on the one hand and excessive cant on the other. Thus most passenger trains ran only a little faster than equilibrium speed. Passengers were comfortable and track damage modest - but so also was the speed.

10. However research has revealed that curving is a more complex phenomenon than had been supposed.

11. As soon as a vehicle standing on a curve starts to move forces develop between the low rail and the leading inner wheel tread which twist the leading wheelset and push it laterally outward so that the outer wheel flange makes contact with the high rail. Paradoxically these forces and therefore also the wearing and derailing tendency, are greatest at low speed. Both the forces and the amount of twist and hence the angle of attack diminish as the speed increases. Therefore, if a vehicle on a curve does not derail in flange climb at low speed, it will be safe at all speeds until it overturns.

12. In consequence there has been a realisation (which has come as a surprise to some) that the determinant of maximum speed on curves is not the physical safety of the train but the environmental needs of the people it carries. This has led to the Sector Director Inter-City challenging the Engineers' judgment and in turn to recent experimentation into perception of comfort and discomfort. Although this work and its analysis are not yet complete it has become plain that whilst previously held opinions about maximum cant maximum cant deficiency and rates of change of cant deficiency were not far off the mark, some changes can perhaps be made in the design of transition curves. Thus whilst on a route such as the West Coast Main Line there seems little likelihood that journey times can be much reduced without recourse to tilting coaches, the next ten years is likely to see some increase in the range of speeds at which trains go round curves, and hence one may expect to see a larger proportion of the track

being run over at full line speed.
Straight Track
13. Can trains on straight track travel faster without
discomfort, and also without unacceptable rates of
deterioration and hence increases in maintenance cost? To
answer this question involves fundamental considerations of
track deterioration behaviour which will be discussed in other
papers at this symposium. It must suffice to observe here that
it is the dynamic increment rather than the quasi-static
element of vehicle loading which produces track damage as a
rule. The dynamic increment is a function of (inter alia) the
initial track roughness and hence the attainment of high speed,
in the absence of radical changes in vehicle design, appears
to present a dilemma. The alternative to increased maintenance
cost seems to be a much higher standard of installation. Even
then some preventive maintenance to arrest even the beginnings
of track movement seem essential. This is the experience of
railways other than B.R. who attempt very high speeds. On such
railways maintenance is done during periods of track possession
which are built in to the timetable. The running of a mixed
traffic railway on a 24-hour basis with the fast passenger
trains running at substantially above 200 kph therefore seem
to the authors are somewhat optimistic expectation about B.R.
track.

MAXIMISING CARRYING CAPACITY
14. Carrying heavy loads is not necessarily synonymous with
supporting heavy axle loads. However, it is a fact of
mechanical engineering life that tractive effort is
proportional to wheel load, and this led to the establishment
of a maximum permitted axle-load for steam locomotives of
22.5 tons on B.R. Gradually over the last 20 30 years, in the
face of road competition, more and more freight traffic on B.R.
has tended to be carried in wagons having axle loads of 23 to
25.4 tonnes instead of about 16 tonnes as heretofore.
15. This level of load remains modest compared with, for
example, axle loads in UK steelworks or on heavy-haul
operations such as those described by Mr. Foley. Nevertheless
it does present the track structure with a load environment
which, taken together with the changed speed pattern
associated with Diesel and Electric traction, is much more
severe than experienced before, say 1960. On B.R. the effects
of this have been: increased rates of rail failure, shortened
sleeper life, and premature collapse of Ballast and subgrade.
To counter these B.R. with assistance from its suppliers has:

- improved its rail steel quality

- extensively employed continuously welded rail

- pioneered concrete sleepers

- increased substantially ballast depth

- developed various methods of subgrade treatment.

These developments will be described in papers which follow. However, it is probably fair to say that they enable the engineer to 'keep up with' the situation rather than looking to a day when heavier still vehicles run, or current loads run faster.

16. The commercial demand for 30 ton axle-load has not been pressed, but assuming that greater axle-load gives an improved payload/total load ratio, then it would be attractive for commercial reasons. The widespread adoption of this, from the track point of view (ignoring the effects on bridges) seems likely to demand at least a stronger rail steel, if not also a heavier rail section.

17. More fundamental work on all the items described will be necessary, but at the same time, track technology might ask 'what is expected from vehicles?' Recent studies show that typical freight vehicle suspensions can have almost any value of stiffness because of variability of friction levels. It is possible to design vehicles which avoid these problems, so giving lower forces on the track, and hence appearing to have lower axle-loads. Similarly, it should be possible to design bogies with lower masses with similar effects.

18. The question of expectations about axle-load poses technical questions some of which only the Civil Engineer is in a position to answer. It also poses 'system' questions relating vehicle first costs to track maintenance costs, which so far have remained unanswered by anyone.

THE LOW REVENUE RAILWAY

19. An inescapable fact of railway life is that large traffic flows along trunk routes are ideal, but steelworks, colleries and oil refineries are not all situated conveniently close to trunk routes. Also, some passenger lines contributing to trunk route flow require reasonable speed capacity, but for relatively small numbers of trains. As the individual revenue contribution of Branch Lines is limited, it follows that their demand on costs must be limited otherwise the branches will whither and eventually the trunk routes will themselves look more like the branches.

20. For freight traffic, the next decade is bound to see development of the (almost) non-maintained railway, where speeds are low, sleepers replaced only when too many in succession are broken or rotten, and rails virtually last for ever. Provided speeds are low, and attention is given to particular features such as sharp curves, there seems no reason why heavy axle-loads could not be contemplated. Here again, we see the commercial influence demanding just how bad track can be before work on maintenance becomes essential (i.e. definition of the worst railway which is acceptable).

21. It is suggested in other papers, better attention to the fundamentals of getting maintenance effectively done, will significantly increase maintenance cycles on fast heavily used lines. If this is so then selective work carried out to a proper standard should lead to significant cost reductions

in the maintenance of minor lines.

22. For passenger traffic on lightly trafficked routes, the first essential is that the Sector Director must be clear about the speeds he requires the trains to run at. Once a sound and unsentimental decision has been made on this point, the Civil Engineer needs to be equally clear and unsentimental about the track standards to be applied. We now have both reliable means for recording track roughness, and reliable advice on how track roughness affects passenger comfort at different speeds. We need to use this information to ensure that maintenance is only applied where it is really needed.

23. It is also becoming clear that on low-revenue lines, whether mainly used for freight or passenger traffic, the traditional B.R. "complete renewal" is a luxury we can no longer afford. Alternative methods are being explored and must be seen as the standard for the future.

MAXIMISATION OF AVAILABILITY

24. Ideally, the railway should be available for use continuously. That this concept is under challenge is due to two factors:

- the development of heavy machinery to replace manual operations in track maintenance and renewal

- the desperate dangers faced by anyone who has to work on or near a track over which very fast trains run.

There is no prospect of reversal of either of these trends.

25. The problems implicit in them have however become more obvious and more severe as a result of several recent developments on the railway:

- route closures

- closure of sidings and loops

- conversion of 4 track sections to double track and of double to single.

- closure of small signalboxes and elimination of associated crossovers.

- greater emphasis on timetabled freight workings at night.

26. It is possible that on routes having the very fastest trains the only ultimate solution may be to do maintenance work during possession, as has been indicated above. In most other places, the power of modern radio communication and the computer are such that it must surely be possible to carry out maintenance involving track possession without actually closing the route. It is recognised that this will demand appropriate timetable scheduling, a limited amount of reversible working, and an adequate provision of facing crossovers.

27. The prizes are considerable:

- maintenance work done during daylight hours and in the week would be cheaper and more effective.

- much less disruption to leisure traffic at weekends.

- increased line capacity of general purpose routes.

28. This seems to be an area requiring more research and one which will require the cooperation of many sectional interests within the railway if solutions are to be found.

CONCLUSIONS

29. The authors approached this paper in a deliberately philosophical way to see if it was possible to derive some general guidelines for future policy in relation to the track. However, it has emerged that "what is expected of the track" is so dependent upon the commercial objectives of the questioner that no single answer is possible.   For the limited range of subjects studied, the following represent our conclusions.

Speed

30. Whilst speeds of 300 kph are technically possible, regular achievement of speeds above about 225 kph over B.R. routes seem unlikely due to problems associated with line capacity, route curvature, and maintenance in a mixed traffic situation. Higher average speeds look possible, and with appropriate attention to fundamentals in both Civil and Mechanical departments may not be all that expensive to achieve.

Load Capacity

31. Whilst heavier axle loads may be commercially desirable, the present situation is more one of keeping up with existing loads. Apart from radical improvement or change in track components, it is suggested that a "system" approach be adopted to consider the possibility of improved bogies to give lower effective loads.

Low Revenue Routes

32. More work is required to define how far track standards can be allowed to decline before work is necessary, and on the necessary control techniques involved. Work is already under way in this area and in the equally important one of economising in materials and techniques for renewal of components.

Availability

33. More flexible working is seen to be needed – with an inter-departmental view of costs and benefits of more reversible working, weekday possessions and timetable adjustments. Progress in this direction would also alleviate problems of high and low speed traffic on one route.

ACKNOWLEGEMENTS

34. In a brief and discursive "state of the art" survey such as this paper, reference is necessarily made to a host of initiatives in hand, and to work already completed and written up. A selection of references on particular aspects is given below, but the authors wish to acknowledge their indebtedness

to colleagues whose work is not mentioned and to their
respective Directors for permission to present the paper.
The opinions it contains however are solely those if its
authors.

REFERENCES

On Vehicle Dynamics   Wickens A.H. & Gilchrist A.O.
'Railway Vehicle Dynamics, The
Emergence of a Practical Theory'.
Council of Engineering Institutions
MacRobert Aware Lecture 1977.

On Curving Behaviour   Elkins J.A. and Gostling R.J. 'A General
Quasi-Static Curving Theory for Railway
Vehicles'. Proc.5th IAVSD symposium on
Vehicle System Dynamics. Vienna, 1977.

On Effects of Rail   Frederick C.O. 'The Effect of Rail
Shape   Straightness on Track Maintenance'
Conference on Advanced Techniques in
Permanent Way Design, Madrid, 1981
(In Spanish).

On Comfort in Curves   Harborough P. 'Passenger Comfort during
High-Speed Curving' Conference on Vehicle
/Track Interaction, Princeton N.J. 1984.

On Lateral Track   Toogood M.J. 'APT/Cl.87 Comparative
Strength   Lateral Forces' BR CM&EE Report 58ID
1980

On Track Deterioration Robson J.D. 'Recent Research into Track
Maintenance Techniques' Conference on
Advanced Techniques in Permanent Way
Design. Madrid 1981 (In Spanish) or
Robson J.D. 'Recent Research into
Track Maintenance Techniques' R & DD
technical memo MRTD 2. 1982.

12   Vertical track loading

C. O. FREDERICK & D. J. ROUND, Research & Development Division, British
Railways Board, Derby, UK

SYNOPSIS. This paper reviews some of the significant
advances made in the understanding of vertical track
loading over the last decade. Areas where further research
would be of benefit to the running of a cost-effective
railway system are highlighted.

INTRODUCTION

1.   The understanding of track loading has improved
greatly in recent years. As train speeds and maximum
axleloads have increased so have the demands placed upon
the track and it has been evident that in many aspects this
increase in demand has only been met economically by the
application of new technology. This is likely to become
even more evident in the future as the operator's desires
for faster trains and more, if not heavier, maximum axle-
loads have to be met and track expenditure has to be minimised.
Track research should suggest more cost effective track
systems, designs and materials and is in fact doing so.
The aim of this paper is to review what has been learnt and
to point towards future developments. With limited space
it has only been possible to consider vertical loads,
lateral loadings bring into play a completely different set
of problems. The nominal axleload of vehicles is only part
of the overall loading pattern as dynamic loads can reach
100% of the static axleload. It is indeed these dynamic
loads which are the principle factors affecting track
design and in general they increase as speeds rise.

2.   It is characteristic of dynamic loads that for any
particular load one must specify a frequency, or duration as
well as a magnitude. In practice it is found that dynamic
loads fall into categories, each of which corresponds to a
group of natural frequencies of the system. The lowest
frequencies are associated with the motion of the vehicle
body on its suspension, there are then in sequence, the
frequencies associated with the motion of the bogie, the
frequency of the unsprung mass on the track resilience and
the frequency of the rail and track components against the
Hertzian spring of the wheel/rail contact and their motion

Track technology. Thomas Telford Ltd, London, 1985

on the flexibility of the rail pads. The spectrum of frequencies can cover the band from O.5 Hz to over 1000 Hz.

3. The track structure is of course designed so that the strongest components experience the highest stresses. The rail experiences the full spectrum of forces generated at the wheel/rail contact. The inertia of the rail and sleepers however tend to resist transmission of the very high frequency forces and little of these higher frequency forces can penetrate to the ballast, even less reaching the sub-grade. It seems likely however that a significant proportion of the middle and particularly lower frequency forces will reach the sub-grade. Therefore when designing track the loading spectrum changes as different layers of the track are considered.

4. The dynamic loads are almost entirely due to irregularities in track profile or stiffness, or irregularities in the roundness of the wheels. The irregularities in the track cause a systematic increase in the loading spectrum seen by particular places along the track, irregularities in wheel profile do not. When considering the fracture of a rail or a concrete sleeper the most severe loading will usually arise due to some combination of the two. When considering the deterioration of track geometry due to ballast compaction or sub-grade settlement then the wheel irregularities are probably not important since they are not systematic.

THE MIDDLE FREQUENCIES

5. These are undoubtedly the most costly to BR and as a consequence much work has been done in this area. Therefore these middle frequencies are considered first. They come about as a result of the dynamic motion of the wheelsets on the track and are significant because of the intimate contact of the wheel and rail. Because the wheels are only separated from the track by the very stiff Hertzian contact spring, the wheel and non-resilient attached mechanisms are referred to as the unsprung mass.

6. An early series of experiments was conducted in order to show the increase in dynamic forces at a rail joint due to an increase in unsprung mass (Ref. 1). The results were surprising since no significant effect was found. In retrospect it is clear why this result was obtained because the rail joint had a level profile and was well packed so that the track irregularity was very small. Later experiments (Ref. 2) were performed with locomotives and dipped rail joints and a clear influence of unsprung mass emerged, establishing both theoretically and experimentally the formula

$$P_2 = P_O + V \propto \sqrt{(MK)}.$$

Where P_O is the static load

V is the vehicle speed

$\propto$ is the angle of the dip

M is the unsprung mass and

K is the track stiffness.

7. Later theoretical studies (Ref. 3) indicated that regardless of the shape of the irregularity the dynamic forces would be a function of MV^2. i.e. if it was desired to increase speed whilst keeping the dynamic loads constant it would be necessary to reduce M so that MV^2 did not change.

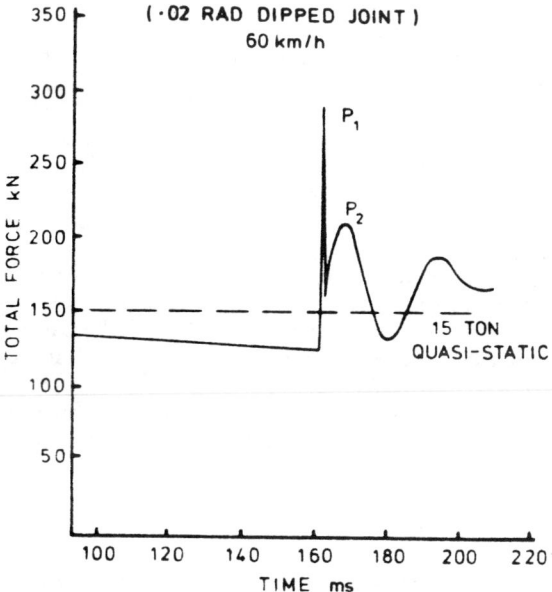

Fig 1. Example of wheel/rail force variation at dipped joint.

8. In pursuing theoretical work on dipped joints and welds, a force was noticed which occurred very near to the running-on rail end and was of very short duration. Subsequent experiments (Ref. 4) confirmed the existence of this force and it was referred to as P_1. (Fig. 1). The magnitude of the force was controlled mainly by the inertia of the rail and fishplates and the ramp angle and only very slightly by the unsprung mass. It was essentially the force due to the collision of the 2 masses i.e. the rails and the wheelset, and since the wheelset mass was much

greater than the effective rail mass, changes in the wheel-set mass had little effect. It is important to realise that on a real dipped joint or dipped rail weld the effective dip or ramp angle controlling P_2 will not be the same as the angle controlling P_1 since the distance over which the P_1 ramp is determined is only a few centimetres whereas the P_2 force is determined by a dip extending over a distance approaching a metre.

9. It was realised that the shear stresses due to P_1 and P_2 at the first bolt hole in the web of the rail acted in opposite directions and thus acted together to increase the range of dynamic stress variation causing fatigue cracks to appear in the bolt hole. Both P_1 and P_2 increased with speed and also increased as their respective ramp angles increased.

10. After a fatigue crack had formed, final fracture was found to be dependent on P_2 alone. Thus the crack size at fracture should become smaller as P_2 increases and it is to be expected that smaller cracks will cause failure on higher speed lines with the unsprung masses and the maintenance condition of the joint also playing a role. This seems to be borne out by experience.

11. The P_2 force is also important in creating high dynamic bending stresses at the welds. The worst conditions seem to arise when the weld is near to a fixed structure. The ballast packing beneath the sleeper is often deficient at these points possibly due to the track stiffness variations on and off the structure and also due to the more complex maintenance operation required. As a result the track support to the rail is weak and thermit welds have been known to fracture as a result of very small fatigue cracks under these conditions (Ref. 5). On BR it has recently been decided that thermit welds on ballasted track should not be nearer than 3 m to a structure where the rails are direct-fastened in order to avoid this problem.

12. On welds which have developed a small region standing proud, due to differential wear, (probably caused by variations in hardness around the weld) there will be an impact force of the P_1 type. This will only be severe at high speeds, but under these conditions it will create high dynamic stresses at the weld and could cause plastic deformation. It is therefore possible that an initially straight weld may eventually become dipped due to P_1 forces after which P_2 forces will develop. Evidence suggests that there is an initial sagging of the weld both when new and after weld straightening, when subjected to traffic. This however stablises after about 6 months. Limited evidence also suggests that this initial sagging may be related to the magnitude of the P_1 force and may be as a result of the internal stresses re-distributing to meet the high dynamic force. This would suggest that any increase in the P_1 force with age could cause the welds to dip further. This long-

term deterioration has not actually been demonstrated experimentally but if it does occur it could provide an incentive to consider the use of a heavier rail section or harder grade of rail steel unless the profile roughness can be reduced.

13. The forces due to the unsprung mass have a very significant effect on the deterioration of track geometry. Evidence, both theoretical and experimental suggests that the P_1 forces are largely filtered out by the rail and sleepers and do not directly affect the ballast settlement (Ref. 6), however the P_2 forces which are in essence the unsprung mass bouncing on the track resilience cause significant deterioration.

14. A great deal is now known about the population of track irregularities as a result of the data produced by the High Speed Track Recording Coach (Ref. 7) and its prototype as well as other systems for measuring weld profiles precisely. The introduction of continuously welded rail on BR has reduced the track roughness by about a half, thereby allowing substantial increases in speed without generating excessive dynamic forces. There is considerable scope for further improvements in this direction by use of modern equipment to accurately control the track construction.

15. Track irregularities develop partly as a result of imperfections in the vehicle but mainly because of imprecise track construction which generate a damaging vehicle response. In recent years there have been efforts to predict the track loading and then go further to predict the increase in track roughness (Ref. 8). In each study however it has been necessary to postulate an initial error in the track construction.

16. If the track is made perfectly smooth with uniform stiffness and uniform ballast settlement characteristics, the dynamic loads are almost zero. The predicted track settlement under the nominal static load is uniform along the track and therefore the track remains perfectly smooth. It cannot be stressed too strongly that tight tolerances in track construction will greatly reduce track loading. Thus the loading on two railways carrying the same traffic at the same speed can be totally different if the track has been built to different levels of precision. An in-depth study of the track quality and maintenance input on 100 miles of the East Coast main line have shown this to be true (Ref. 9). Work is continuing to define tolerances for the various aspects of construction.

17. One of the important sources of track irregularity is that associated with the longitudinal profile of the rail. It is important to know how initial input errors affect the subsequent track profile. A simple model of track behaviour has been developed which shows how poor tolerances in the rails lead to deterioration of track geometry and how the vehicle loadings interact with the initial construction error

to cause track roughness (Ref. 10). The model is also able
to consider the effect of irregular ballast profiles on the
subsequent track profile.

18. In the model the rail is supported continuously along
its length by ballast and the settlement characteristics of
the ballast are matched so that a specified load on the
track will give a settlement rate equal to that which would
be experienced by a real track with sleepers i.e. periodic
support. The vertical resilience of the continuous support
is also matched to that of real track. Into this model it
is possible then to introduce rails of irregular shape and
also an irregular longitudinal ballast profile. Usually the
fitting together of the rail and ballast profiles produces
and uneven distribution of force on the ballast and an initial
irregular track profile. A vehicle is then run over the
irregular track profile and the dynamic wheel/rail forces
are calculated using a set of linear vehicle and track
dynamic characteristics. This theoretical vehicle may be a
single unsprung mass, a 2 axle vehicle or a 4 axle bogie
vehicle. The settlement of the track at each point is
determined by the combined effect of the static axleload,
the dynamic axleload at that point and the initial uneven
force distribution due to the lack of fit between the rail
shape and ballast profile. The settlement relationship is
assumed to be linear i.e. doubling the axleload will double
the settlement rate for a specified number of axle passes.
All the differential equations in this model are linear so
that it is possible to add together the effect of different
rail or ballast irregularities.

19. Because of the linearity of the model it is possible
to sub-divide the rail shape or ballast profile into
Fourier harmonics of various wavelengths, to calculate the
individual effects on track geometry and then add the
individual effects in order to see the overall results. It
is also possible to define a transfer function which shows
the effect on track profile of rail or ballast profile
irregularities at various wavelengths for a specified number
of axle passes. The transfer of rail irregularities into
the track profile with quasi-static loading and how these
develop with the passage of traffic is shown in Fig. 2.

20. However for calculations involving significant
dynamic forces it was found to be necessary to include a
further stage in the calculation. The initial model
assumed that the maximum load experienced by the ballast at
any particular point along the track occurred when a wheel
was directly above the particular point. For impact loads
of the P_2 type this is not always the case.

21. The deflection pattern of rail under a P_2 force is
essentially the same as that under a static force and using
well known formulae for a beam on continuous elastic support,
the deflection of the rail defines a pressure distribution
on the ballast. To find the maximum pressure experienced

by the ballast at any point during the passage of a wheel it
is necessary to find the envelope of this pressure distribution as the wheel moves along the track.

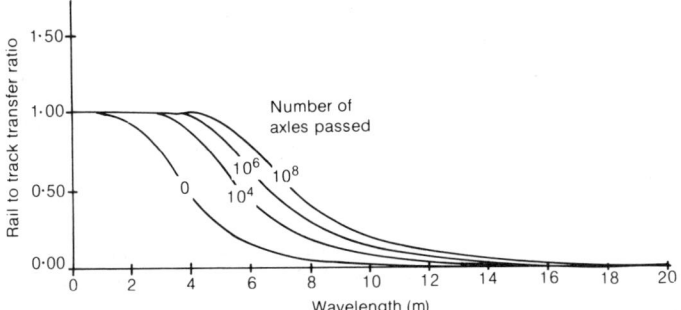

Fig 2. Theoretical transfer of component
wavelengths of rail shape due to
quasi-static loads.

22. If the dynamic forces are small compared to the static
axleload, the envelope of ballast pressure distribution will
be identical to that obtained by assuming that maximum
forces occur under the wheelsets. If the dynamic forces are
large compared to the static axleload some points will
experience the highest pressure when the axle is nearby and
not directly overhead.

23. In a linear model the dynamic forces due to the unsprung
mass must average out to zero along the track i.e. a downward
impact force at one point must be balanced by an unloading at
adjacent points. The frequency of the P_2 force is typically
about 50 Hz, so at 50 m per second one cycle of oscillation
occurs within a distance of 1 m. Over such a short distance
the rail can easily bridge the effects of loading and
unloading so that the nett effect of the P_2 force on the track
settlement is very small. When the envelope effect is
included however it is equivalent to a superimposed load
distributed along the track and equal in magnitude to the
difference between the envelope of pressure and the pressure
derived by assuming the wheel is overhead. This superimposed
load can itself be included in the linear model to give an
approximate solution. This correction produced a marked
increase in the predicted track roughness and created a
situation where severe short wave irregularities can generate
track irregularities at longer wavelengths. In more
comprehensive track models which realistically model sleeper
spacings this envelope effect is somewhat obscured, but the
general characteristics of behaviour which are predicted
are very similar to those shown by the simple model.

24. Probably the most significant error in the simple model
lies in the assumed settlement behaviour which is of time-
hardening type, since it is known that ballast behaves much
more like a strain-hardening material i.e. settlement rate

depends upon load and the settlement already experienced. However the simple model does illustrate some interesting effects concerning track loading and it is probably reasonably accurate for a uniform traffic type in which the dynamic loads do not change markedly with the number of axle passes. The validity of the simple model has been checked against more complex models of specific aspects of track of vehicle behaviour in doubtful cases.

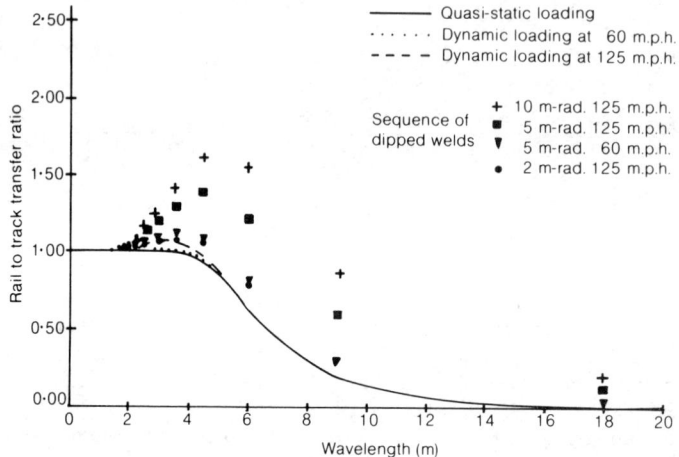

Fig 3.   Theoretical transfer of component wavelengths of rail shape including dynamic effects of the unsprung mass.

25. Fig. 3 shows the track irregularities resulting from rail irregularities, as a result of the unsprung mass interaction. It should be noted however that the points showing enhancement of the harmonics of the rail length come not from rail faults at this wavelength but from the shorter length faults in the rails as described above and Fig. 3 is therefore not strictly a transfer function.

26. Experimental measurements (Ref. 11) on the transfer of rail irregularities to the track profile, indicated that there were features which could not be explained by the unsprung mass effects alone.

LOW FREQUENCY LOADINGS

27. From the work described above, it was clear that significant enhancement of, most noticeably, 9 m wavelengths was occurring which could not be explained by unsprung mass effects. This led to vehicle body and bogie effects being examined. It was suspected that as the rail was flexible at these long wavelengths it would not be able to support even small loads associated with bogie or body pitching or bouncing. Results however suggested that if vehicles were made to design specification they should have little effect on the track. However if the damping was insufficient to curtail the

natural frequency response of the vehicle to forced
frequencies related to the rail length then significant
deterioration would occur. Theoretical calculations of this
effect on a measured rail gave track profiles typical of those
found in reality. An example is shown in Fig. 4 with the
real profile resulting from this rail shown in Fig. 5. It is
clear that the track profile predicted to result from the
combination of unsprung mass and body pitching is very similar
in character to the actual track profile.

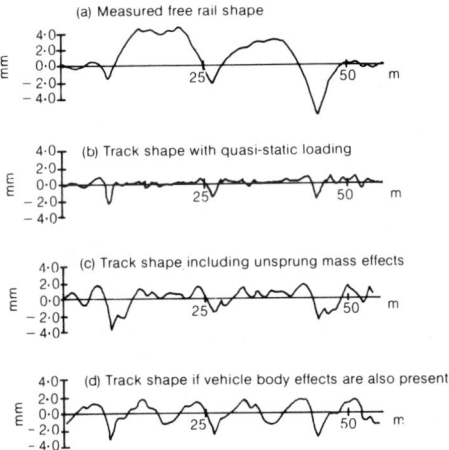

Fig 4. Track profiles theoretically predicted to result
from a section of measured rail (a) due to quasi-
static loading (b), unsprung mass effects (c) and
vehicle effects (d).

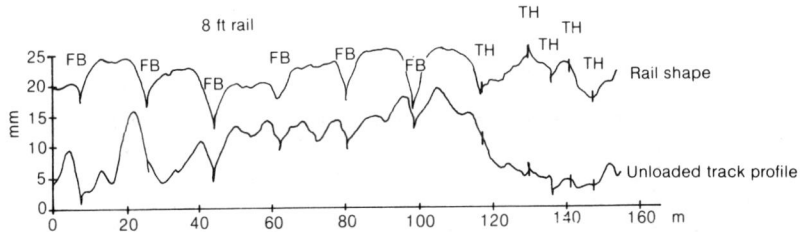

Fig 5. Measured rail shape and resulting measured track
profile (FB = Flash Butt Weld, TH = Thermit Weld).

28. These long wavelength enhancements would most likely
be the results of 4 wheel freight vehicle pitching or of
loco bogie pitching both of which occur typically near their
normal working speed range, 35 to 50 mph and 80 to 110 mph
respectively depending upon the vehicle type. Both manifest

themselves similarly in the track and hence only one example
is shown in Fig. 4. Vehicles will be examined for significant
evidence of these phenomena in the near future, in an attempt
to isolate the worst offenders.

29. Most experiments to measure P_2 forces have been made
with locomotives since these usually exert the highest dynamic
loads. However high loads can also be exerted by heavy
freight wagons. In tests to compare the impact load of 25 ton
axleload freight wagons it has sometimes been found that the
trailing axle of the bogie exerts higher forces than the
leading axle. This has not yet been fully explained
satisfactorily but could be significant. Fig. 6 shows how the
peak baseplate force varied with speed for leading and
trailing axles. An appropriate conversion of peak baseplate
force to axleload is to multiply by 4.(See Discussion, reply
to Mr Mau, question (c).)

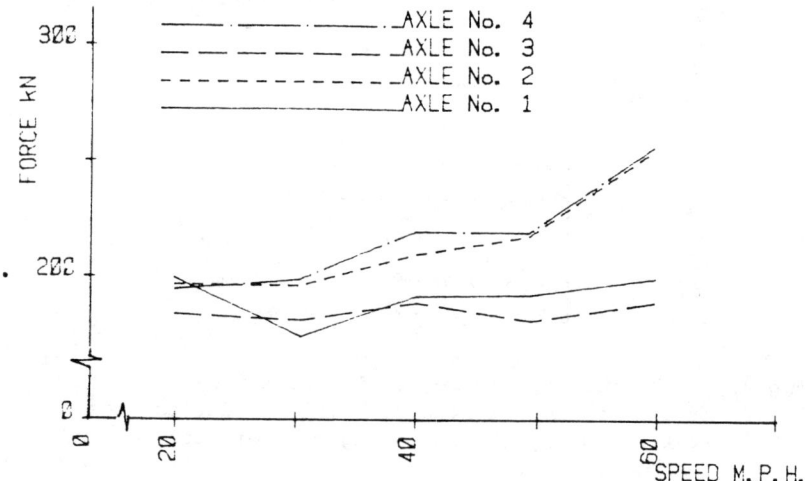

Fig 6. Peak baseplate loads due to passing of 100 ton tank
wagon showing characteristic tendency of trailing
axle to give higher forces. (Axles 2 & 4 are
trailing).

30. It has also been found that friction damped vehicles
have dampers in which the breakout force increases with age.
If this breakout force becomes too large these vehicles
could significantly affect the track deterioration since the
effective unsprung mass becomes very large. A rig has
recently been developed to enable vehicle unsprung masses to
be measured, and it is planned to measure the effective
unsprung masses on some friction damped vehicles in various
states of maintenance sometime in the near future.

HIGH FREQUENCY LOADINGS

31. The most severe phenomena of this type on BR are
associated with short pitch rail corrugations (40 to 70 mm

wavelength). Since the forces are periodic, they can excite resonant responses in the track components. This response is particularly serious for concrete sleepers (Ref. 12) since it can encourage cracking or loosening of the inserts which hold the rail clips. Fig. 7 shows measured dynamic responses in concrete sleepers and other parts of the tracks due to corrugation. This resonant response is in addition to the quasi-static stresses and stresses due to the P_2 effects, but the amplitude of the resonant response will be more heavily damped when a sleeper is carrying a large load. As yet there have been no experiments to measure strains in concrete sleepers under a combination of P_2 forces and corrugations.

32. It is sometimes not appreciated how shallow is the depth of rail corrugation. A depth of 0.1 mm is typical. This depth however is sufficient when travelling at speed to cause very large variations in the wheel/rail force because the contact spring between the wheel and rail is so stiff. The differential wear at a weld can lead to the weld zone being raised by 0.08 mm. This roughness creates dynamic forces at high frequencies which become significant as speeds rise even if the excitation covers a wide band of frequencies. These forces are significant partly because they generate noise but also because they raise the wheel/rail contact stresses and thereby create a fatigue problem. In the case of corrugations, rails are found with the longest and most numerous fatigue cracks near to the crests and in the case of welds it is common to find squat defects on the running surface.

33. It is known from laboratory studies (Ref. 13) that rolling contact fatigue worsens dramatically as the contact

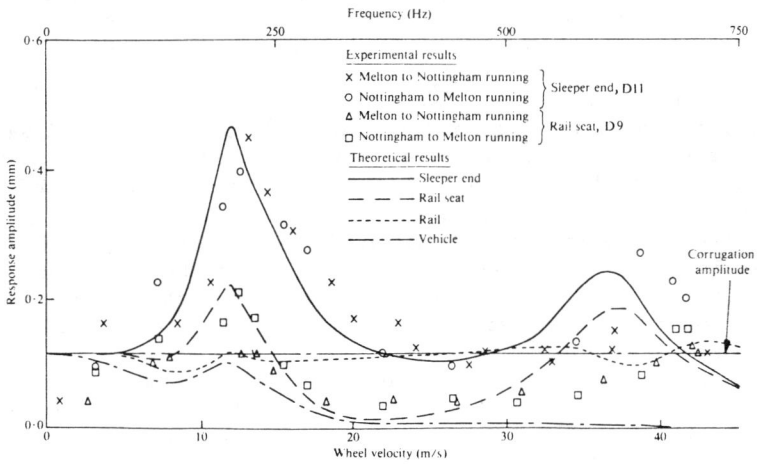

Fig 7. Dynamic response of track to vehicle passing over corrugated rail.

stress increases and is also sensitive to high values of creepage between the wheel and rail. In general, assuming the wheels are made of steel,

$$\text{Contact Stress} \quad \propto \quad \left(\frac{\text{Contact Load}}{\text{Transverse Rail Wheel Radius}} \right)^{1/3}$$
$$\text{Radius}$$

34. Both P_1 and P_2 forces can produce marked increases in the contact load, so both these quantities should be considered when comparing contact stress situations. These forces rise with speed, therefore it is clear that at higher speeds the maximum contact stresses will be higher and the fatigue life of the rail shorter.

35. The use of locomotives with a large tractive effort can provide a further worsening of the situation if they are allowed to operate at high values of creepage. A great deal remains to be discovered concerning the effect of traction and braking in causing fatigue of the rail surface. For the present it would seem sensible to develop locomotive control systems which prevent high values of creepage and to grind the surface of the rail to reduce P_1 forces and remove damaged material.

36. The dynamic stresses in rails and concrete sleepers due to wheelflats have been studied extensively (Ref. 14 & 15). It has been shown that the impact causes bending waves which propagate along the rail at high speed and cause tensile stresses in the head of the rail sufficient to cause a fracture when there is a pre-existing defect. Most of this work was done using artificially created wheelflats or equivalent indentations in the head of the rail. It is now suspected that this may not be a true representation since in a real wheelflat, material may be moved from one place to another as well as being removed altogether. A projection on a wheel could create a more severe effect than a simple flat. However the work which has been done shows the existence of 2 peaks in the curve of dynamic load against running speed (Fig. 8). The first corresponds to oscillations of the unsprung mass on the track spring and occurs at very low speed, the second corresponds to the impact between the wheel and the rail which occurs after the wheel has separated from the rail as it "flies" over the flats. Rail fracture problems are more likely to be associated with this second peak. The severity of the impact is governed by the extent to which the rail was pressed down on the track resilience before being released by the flat, and also depends on the flat being large enough or the vehicle travelling fast enough to ensure that the wheel "flies".
The impact at high speeds is largely caused by the motion of the rail. To obtain a good correlation between the theory and an experiment in which the rail strains were measured it was necessary to include both bending and shear deformation of the rail in the model (Ref. 14).

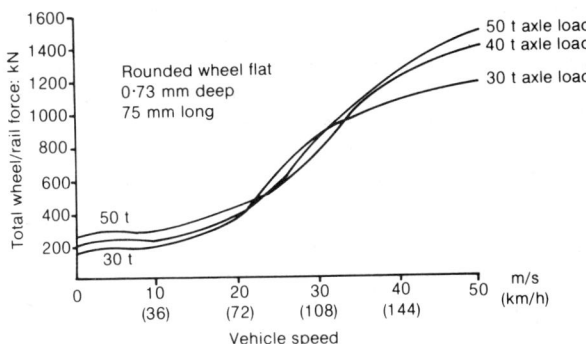

Fig 8. Calculated forces due to wheel-flat impact
superimposed on P_2 type forces.

37. In the USA strain measurements on concrete sleepers in
service revealed the presence of long flats up to 400 mm long
possibly caused during wheel turning. These created very
high stresses in the sleepers. To date the possibility of
such flats on BR has not been investigated but it is
potentially a very important question since only a small
number would be needed in order to significantly worsen the
dynamic environment of concrete sleepers in the track. One
such axle could be equivalent to dipped weld operating every
3 m along the track.

DISCUSSION
38. As more is discovered about track loading it becomes
ever clearer that greater efforts must be made to build track
to a higher precision. The ballast bed profile must be smooth
and uniform, the rails must be straight and the running
surface of the rail must be smooth. This precision must be
provided at all wavelengths affecting track loads. For a high
speed railway this means from a few centimetres to many
metres. At the shortest wavelength, amplitudes of a few microns
are significant, while at the longest wavelength, amplitudes of
1 or 2 mm become significant. Not only must the track be
built correctly, the rail surface must also be periodically
smoothed. The tolerances required for many components have
investigated but others remain unknown.

39. There is a clear need to learn more about the roughness
of the circumferential wheel profiles since these could be
important in track loading. There is also a need to learn
more about the rolling contact fatigue of rails and the
extent to which it is affected by traction and braking.
Finally an effort should be made to see whether rail welds do
in fact deform in service and if so what can be done to
prevent it.

40. There are several indications that more should be
learnt about the way that suspensions of freight vehicles and
locomotive bogies behave in service. The deterioration of

track geometry could be greatly worsened by incorrect suspension performance. There have been reports that some vehicles which rely on friction damping require a significant force to overcome the friction. This may reveal the reason for trailing axles exerting a higher impact load than leading axles on the same bogie, and why some vehicles are worse than others in generating ground vibrations (Ref. 16).

ACKNOWLEDGEMENT

The authors wish to thank the British Rail Board for permission to publish this paper and to acknowledge the efforts of many people too numerous to mention individually who have contributed to the various fields of research described in this paper.

REFERENCES

1. LOACH J.C. (1965) "Research into some factors which influence the vertical loading of railway track" Paper No. 6823 Proc. Inst. Civ. Engrs. Vol 30 pp 731-746.

2. NIELD B.J., GOODWIN W.H. (1969) "Dynamic loading at rail joints", Railway Gazette, 15 Aug pp 616.

3. FREDERICK C.O., "The effect of wheel and rail irregularities on the track" 1st International Heavy Haul Conference Perth, Australia Sept 1978.

4. JENKINS H.H., STEPHENSON J.E., CLAYTON G.A., MORLAND G.W., and LYON D. (1974) "The effect of track and vehicle parameters on wheel/rail vertical dynamic forces". Railway Engineering Journal, Vol 3 No. 1.

5. FREDERICK C.O., and ASHTON M.E., "Derailment at River Wharfe Bridge, Ulleskelf, E.R. on 8 Dec 1981 R. & D.D. investigation" British Rail Research & Development Division Tech. Memorandum Nov 82.

6. ROUND D.J. "Tolerances for the straightness of C.W.R. to minimise track maintenance" British Rail Research & Development Division Report March 1984.

7. High Speed Track Recording Coach Handbook.

8. LANE G.S., "The effects of track and traffic parameters on the development of track vertical roughness". 2nd International Heavy Haul Conf. Colorado Springs, U.S.A. 1982.

9. ROUND D.J., "The East Coast Main Line Maintenance Project – first report" British Rail Research & Development Division Report March 1984.

10. FREDERICK C.O. & ROUND D.J. to be published.

11. ROUND D.J. "The effect of rail straightness on track geometry and deterioration", Rail Technology - proceedings of a joint AAR/BR seminar published 1983.

12. CLARK R.A., DEAN P.A., ELKINS J.A. and NEWTON S.G. "An investigation into the dynamic effects of railway vehicles running on corrugated rails" Jrnl Mech. Eng. Sci. 1982.

13. CLAYTON P., ALLERY M.B.P. & BOLTON P.J., "Surface damage of rails" Rail Technology - proceedings of joint

AAR/BR seminar published 1983.

14. NEWTON S.G. and CLARK R.A. "An investigation into the dynamic effects on the track of wheel flats on railway vehicles", Jrnl Mech. Eng. Sci. 1979, 21, 287-291.

15. CANNON D.F. and SHARPE K.A. "Wheel flats and rail fracture" Rail Technology - proceedings of a joint AA/BR seminar published 1983.

16. DAWN T.M. and STANWORTH C.G., "Ground vibrations from passing trains" Jrnl Sound & Vib. 1979 Vol. 66 No. 3.

Discussion on Papers 10–12

MR S. T. LAMSON, Research Associate, Canadian Institute of
Guided Ground Transport, Kingston

In the use of a tamping index for determining tamping
priority, it is presumed that the index reflects the rate of
track geometry deterioration or (on a more complete basis) the
cost of track and rolling stock maintenance. Dr Esveld, is
this the basis for the choice of formula for the tamping index
presented in the paper?
   The MISO concept is a neat way of processing the vast
quantity of track geometry data into a rational track quality
index or set of indexes. From a similar effort at the
Canadian Institute of Guided Ground Transport on the vehicle
response to a range of wavelengths the track geometry error is
too complex to be satisfactorily estimated with a single
transfer function. Could you elaborate further on the details
of the transfer functions in MISO? Do you use a single
function or a set of functions for each range of wavelengths?

DR C. ESVELD

The tamping index for determining the tamping priority
reflects the quality of the track geometry in the waveband 0–
25 m for a maintenance section of 5–10 km length. If
sufficient historical data have been collected the tamping
index may be extrapolated to, for instance, half a year ahead.
The formulae governing the relationships between standard
deviation and quality index are based on the national
distributions obtained during the recording car run of spring
1983.
   This approach is purely empirical, without physical
substantiation. The impact of maintenance reduction on
passenger comfort, track loads, safety and costs cannot be
answered in this way because the link to vehicle reaction is
missing.

A more detailed discussion on the assessment of track quality is given in the appendix to my paper.

Nederlandse Spoorwegen determine the necessary relations between track geometry and vehicle response, called transfer functions, with the MISO method. In principle four track geometry components are considered, i.e. cant, level, alignment and gauge, as illustrated in Fig. 7. For the on-line calculation of vehicle response the transfer functions are approximated by recursive filters. This approach has the great advantage of substantially reducing the computational time compared with ordinary convolution. A study is presently being conducted to see whether these filters should be made from microprocessors or analog filters. For further details on the MISO method please see reference 3 given in the paper.

MR A. COOKSEY, Assistant Inspecting Officer of Railways, Department of Transport, London

Mr Cope and Mr Gostling, you suggest that on some freight-only 'low revenue routes' standards should be allowed to fall and, perhaps, to allow failure to occur before maintenance work is undertaken. What do you see as the risks of such non-maintenance?

You also suggest that a control technique will have to be devised. What form should this take? Presumably both high technology systems and traditional manual inspection would be equally unacceptably costly.

MR R. J. GOSTLING

The phrase used in the paper was the (almost) non-maintained railway rather than the suggestion that maintenance should only follow failure. Experience in other countries shows that trains are able to travel - albeit slowly - over tracks which we might hesitate to use as sidings. Although the situation on British Rail could be different in some respects, e.g. the greater use of two-axle vehicles compared with the American practice, some change in this direction is suggested.

On low traffic routes the deterioration, and hence the requirement for maintenance, which results directly from the passage of trains, is likely to be very small, with problems from the effects of weather and time assuming greater importance. In terms of safety, the first limit to be reached would probably be the capacity to hold gauge - hence the suggestion that a minimum level of maintenance would include some spot re-sleepering. A comprehension of where the largest forces will be generated (e.g. in sharp curves) would help to guide this policy, allowing reduced maintenance with little effect on derailment risk.

MR P. G. RAWLING, Area Civil Engineer, British Rail, Sheffield

In paragraph 13, Mr Cope and Mr Gostling mention that the alternative to increased maintenance costs is a much higher standard of track installation. British Rail has recently tried to achieve such improved installation on the new section of the east coast main line constructed to avoid the new Selby coalfield.

The feature of interest is the compaction of the ballast bed by the Boschung vibrating multi-plate compactor mounted on a Mercedes Unimog vehicle. Four passes were normally made in front of the track re-laying.

Guide rails to a 10 ft gauge were then pulled out along the ballast bed from a CWR train using the same Mercedes vehicle.

Normal track re-laying was done by a Pluto sleeper bale gantry running on the 10 ft gauge guide rails. When sleepers had been laid, the guide rails were transposed to become the running rails using the KARP rail positioner. Clipping up and welding then produced the final track.

As laid this had a remarkably consistent standard of quality giving just over 2000 Neptune penalty points per quarter mile. (The Neptune track recording system was used with the Matisa recorder because the high speed track recording car could not run on the unfinished railway.)

Ballasting up and one pass of the tamper giving 15-20 mm lift improved the standard to 1000 penalty points.

A further pass giving a lift of 10-15 mm further improved the standard to between 200 and 300 penalty points per quarter mile.

With this finished standard we hope that the track will have a good start and will give a good ride at 125 m.p.h. The track was installed at about half the cost of normal track renewal - but under 'green field' conditions. The line has been opened for the first trains at 70 m.p.h.

MR R. J. GOSTLING

It is encouraging to hear of the efforts of engineers in the field to achieve high quality track installation. Perhaps the most important task is to define a set of standards which are partially achievable and would produce good long-lasting track. Work at Derby has helped to clarify these standards and the specification for the welding depot at Redbridge includes a requirement to produce all welds with an angle of less than 2 moad, measured over a 1 m straight edge. Such quality is now regularly achieved on other administrations, with many welds significantly better, and this contributes greatly to better and less costly track. The required accuracy of ballast bed is less precisely known at present. However, a sinusoidal irregularity of 6 mm peak to peak would produce by itself a standard deviation of 2.1 mm - the '90%

better than' criterion used by the Chief Civil Engineer. As
other contributions to roughness exist, and new track should
aim at a standard deviation of half this value, it is evident
that the compacted ballast surface requires a level variation
of less than 2 mm. This would be required over a length of
20 m or more.

MR J. BUEKETT, Director, Costain Concrete Co Ltd, Hoddesdon

The Canadian National Railways route through the Rocky
Mountains is a heavy haul railway with most traffic consisting
of waggons with 30 t axles and 40-45 million gross tonnes per
year on the single track.

In the early 1970s a decision was taken on this route to
install concrete sleepers in all curves tighter than 450 m
(later increased to 900 m) and where there is less than
0.5 mile of straight track between curves. The concrete-
sleepered track compares successfully with the previous soft-
wood-sleepered track. The rail life has increased, there are
extended intervals between relining and relevelling, a much
reduced derailment frequency, an improved ride and a reduced
fuel consumption, i.e. an increased initial track cost has
resulted in a saving in overall track costs.

This track has 68 kg/m rail and 610 mm sleeper spacing,
thus giving a greater stiffness than typical British Rail
track, and the trains are relatively slow at 30-40 m.p.h.

Mr Gostling, you have expressed concern about the
accelerated track deterioration caused by freight vehicles
with 25 t axles. Is the effect of axle loads so serious and,
following Mr Bonnett's earlier comments, are you sure that a
large number of light axles is preferable to a smaller number
of heavier axles?

You have shown that the rate of track deterioration is
proportional to the speed, but I suggest that a power function
of speed relates more closely to what is observed in track.

Could you comment on the relationship between track
stiffness and deterioration?

MR R. J. GOSTLING

Work undertaken on ballast settlement supports the view that a
small number of heavy axle loads is equivalent to a larger
number of light ones. There is some dispute on the numerical
relationship, with one school of thought even suggesting that
lighter axles are not relevant at all in the final
deterioration. Work is actively in hand at Derby with a half-
sleeper rig capable of programmed loading to simulate traffic
patterns to clarify the situation.

The relationship with speed, represented as a simple
linear formula, is supported by studies using the track
deterioration program. Calculations to estimate the effects

of speed changes suggest a relationship which is rather less
than linear, but this depends on the percentage of random
components in the behaviour.

With regard to track stiffness, it is apparent that the
deterioration rate increases with both higher and lower values
of stiffness. At high values, the dynamic forces experienced
will increase, leading to greater settlement of the ballast.
At lower values, deterioration is likely to be caused by
excessive movement of the ballast particles, resulting in
higher rates of degradation.

MR K. G. A. HERRING, Area Civil Engineer, British Rail,
Newcastle upon Tyne

I would like briefly to develop the aspect of the re-laying
and maintenance requirements of various classifications of
track. Mr Cope and other speakers have put forward the
principle of the importance of a high standard of installation
for track re-laying of high speed, heavily used railways which
will then require less maintenance. This principle has been
proved in practice. Reference has also been made by Mr
Gostling and others to the importance of a high standard for
formation and ballasting work. I have had the opportunity
when carrying out mid-week and weekend track re-laying work to
consolidate the bottom ballast before laying the track panels
and have found this to be very effective. It has provided a
much better track standard after re-laying and the track has
required the minimum of subsequent maintenance for several
years afterwards. Another advantage is that I have been able
to hand the track back to the operating department at a higher
level of speed restriction which has assisted me in obtaining
better track possession and speed restriction facilities for
other work on the same route.

There is also a need to carry out minimum treatment by
resleepering only on low speed freight-only railways but I
consider a low cost method of re-laying should be investigated
as a viable alternative for medium traffic density routes.

My problem in North East England is that on many lines as
well as decaying sleepers there is no ballast, which causes
maintenance difficulties and in these cases I am not sure that
just to change the sleepers is the complete solution.

I would welcome comments on the introduction of a policy
for some routes of inexpensive re-laying by

(a) the reuse of existing rails including welding the
existing rails on site if possible
(b) complete resleepering using a sleeper bale method to
remove the old sleepers and to replace them using
serviceable refurbished sleepers
(c) minimum reballasting to a depth of 150 mm under the
sleeper of if clearances allow by lifting the track on
top of the existing formation (this minimum ballast depth

would then enable the track to be mechanically
maintained)

(d)   consolidation of the ballast and the provision of a good
top and line to the track before handing back to traffic
to reduce subsequent track maintenance costs

(e)   the introduction and development of improved single-line
methods of executing re-laying work which will enable re-
laying to be carried out economically with the minimum of
traffic disruption during normal weekday shifts (this
aspect has considerable potential for cost savings and
should be discussed more fully as a separate subject).

This low cost solution to the problems of the medium
category freight-only lines could, I suggest, be considered
and may prove to be more cost effective in the long term.

MR R. J. GOSTLING

Mr Herring's comments are welcome as they provide practical
back-up to the view that good quality relaying leads to long
life on heavily used lines.

The replies to his particular questions depend very much
on the line categories being considered. For routes with
relatively infrequent traffic I suggest that considerations of
income preclude almost any development to a better track
quality. The traffic on such routes does not justify the
expenditure, in today's economic climate, however unattractive
this idea sounds to anyone attempting to do a good job.

However, if secondary through routes with a reasonable
amount of traffic are considered, many new ideas are worthy of
careful consideration. The concept of more use of single-line
working to allow more economical weekday relaying and the
optimum re-use of serviceable materials could be useful and
the concept of achieving better 'top' to provide longer life
has already been dealt with.

In summary, it may be that Mr Herring's 'cut-price'
suggestions are still too expensive for the type of lines he
mentions, but that a more 'up-market' application could be
possible.

MR R. MITCHELL, District Engineer, Australian National
Railways, Invermay

Mr Cope and Mr Gostling: is it universally true that a train
is safer at higher speeds on curves until overturning is a
problem or, if not, within what bounds is it true? Is
increasing curve speed a solution to flange climb derailments?

In the Tasmania region of the Australian National
Railways we have 100 m (5 chain) curves on a 1.067 m
(3 ft 6 in) gauge. The track is mainly short rail and under
ballasted. Irregularities on the top are therefore frequent

and sometimes excessive. Our approach has been to keep curve speeds down on cant deficiency criteria but we still have derailments in flange climb. Would you like to comment on this?

MR R. J. GOSTLING

It is generally true that, unless it significantly overspeeds, a train is safer on curves at higher speeds than lower. At lower speeds, derailment on curves is normally caused by creep forces acting between the wheel and the rail, whereas at high speeds the lateral forces arise from centrifugal effects. With an increase in speed, the tendency is for derailment risk from creepage forces to decrease and it is not until of the order of 20% of cant deficiency (depending on vehicle height, track gauge etc.) is achieved that derailment occurs. Thus there is a very wide central margin in which derailment is unlikely but passenger comfort is the key issue.

The case quoted, of sharp curves with poor track top, is a classic example of derailment-prone track. The track twists can redistribute vertical load to increase the lateral creep-induced forces on the low rail and to reduce the vertical restraining force on the high rail. Limiting curving speed is not a particularly effective solution, although it minimizes resulting damage. An interesting remedy is to lubricate the head of the low rail with some non-persistent substance – dilute detergent for example. Another possibility is to pay particular attention to those twist faults which cause unloading of the outer wheel, i.e. excessive dips in the high rail, since these are much more likely to lead to problems.

MR M. CHORLEY, Area Civil Engineer, British Rail, Perth

I shall use Mr Cope's remarks on the problems of the Dingwall-Wick railway line as an introduction to our alternative re-laying strategy for railway lines beyond Inverness together with the current progress and costs.

The railways from Inverness to Kyle of Lochalsh (westwards) and to Wick/Thurso (northwards) consist of 234 miles of single track – bull head rail, fishplated joints and timber sleepers – mostly 25-45 years old. As Mr Cope stated, conventional re-laying costs approximately £200 000 per single track mile and could not be justified by traffic revenue available in this sparsely peopled region.

So what can we do and what are we doing about track costs?

(a) Rails
A bad engineer changes rails after they break in the track, a poor engineer changes rails long before they would have broken in the track, but a good engineer changes rails

just before they would have broken in the track.

Rails break because of fatigue (load plus frequency).
There are three trains per day beyond Inverness of 18 ton
maximum axle weight. Therefore British Rail have allowed an
increased wear of 2 mm of overall depth of rail before re-
laying is required. In the unpolluted atmosphere and light
traffic this reduction gives another 20 years life to the
rails and also allows side-worn rails on curves to be replaced
with bull head rail cascaded from re-laying of more heavily
trafficked routes at the cost of labour only.

(b) Sleepers

Controlled patch resleepering does not mean resleepering
every third sleeper but only changing rotten sleepers
identified by careful examination. This may be as many as ten
in one 60 ft length of track or as few as two in another
section. A maximum of 30% resleepering provides reconditioned
track capable of another 15 years life without further major
expenditure.

(c) Rail joints

Fishplated joints are costly to maintain and become a
potential point of breakage at the rail bolt holes. We have
now started to cut off the ends of rails containing the old
fish bolt holes and to weld in situ to produce serviceable
bull head continuously welded track.

(d)   Relative costs (all per single track mile)

Conventional re-laying: serviceable continuously welded rail
      New concrete sleepers   £190 000
      Serviceable concrete sleepers   £120 000

Controlled patch resleepering (new locally
      grown timber)   £35 000
      plus rerailing (when required on curves)   £8 000
      plus cropping rail ends and welding   £7 000
                                                       _____
                                                       £50 000

Even the most expensive combination of controlled
patching is £70 000 less than the cheapest complete re-lay.
However, bull head timber track costs about £1000–£1500 more
in labour to maintain than flat-bottom continuously welded
rail and concrete sleeper.

The moral is 'you cannot eat your cake and have it'.

(e) Conclusion

Twelve months ago trains beyond Inverness were empty and
three bus operators were supreme between Inverness and Wick.
A series of radical cost cutting and promotional initiatives
by all railway departments in the area business group at
Inverness has reversed the situation.

One coach operator is bankrupt, the second has withdrawn from competition and the third has dismissed 14 drivers and has reduced his frequency of service. The Highland and Island Development Board has decided to pay for the painting of all 33 stations on the Kyle and Wick routes. The trains are full again and the future of the railway lines in North Scotland is good, and the railway looks secure to remain as a public service into the 21st century.

DR A. ZAREMBSKI, Director of Research & Development, Pandrol Incorporated (USA)

Regarding modelling of vehicle-track interactions on corrugated track in the USA involving North American freight corrugations with wavelengths 100 mm $<\lambda<$ 300 mm at speeds of 20–50 m.p.h. and depths of 25-2 mm we have found

(a) very short lengths or corrugations (as little as 4-6 wavelengths) are needed to reach peak excitation
(b) dynamic impact factors of 2 or greater can be achieved at these speeds and depths
(c) a significant level of fuel consumption due to energy loss in the vehicle suspension (see Fig. 1).

In your modelling or test work, Mr Frederick, have you noted similar effects for the very short wavelength corrugations that you have investigated?

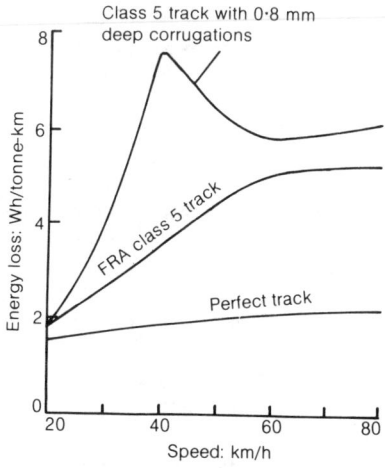

Fig. 1. Energy dissipation for perfectly straight and smooth track compared with FRA class 5 track with and without 0.8 mm deep corrugations

159

MR C. O. FREDERICK

It is interesting to note that Dr Zarembski has observed
significant energy absorption due to long wave corrugations
and that this probably relates to increased motion of the
suspension. On British Rail our corrugations are very largely
short wavelength. We have not sought to measure the energy
loss due to these corrugations but it is likely that if there
is a significant effect it will be associated with enhanced
motion of the track components since the frequencies are too
high to allow significant motion of the wheelset or
suspension. Our knowledge of dynamic effects due to
corrugations shows that dynamic impact factors of 2 are
possible and factors higher than this can occur in concrete
sleeper strains due to resonance effects.
    The damping in the sleepers ahead of the wheel is
presumably very small as the precession wave takes effect.
Thus only a few oscillations would be expected to be required.
Also, the vibrations must change under the actual wheel
passage, when resonant oscillations will be more heavily
damped by the ballast. Thus the pattern must build up quickly
to achieve peak values before a reduction under the wheel
itself.
    The experiments did not include measurement within a very
few cycles of the start of the corrugated length, so an
experimental answer to the question is not available.

MR F. I. MAU, Civil Design Engineer, BHP Engineering, Sydney

Mr Frederick, please could you explain the following.

(a) Paragraph 9 – why do the shear stresses due to P_1 and P_2
    at the first bolt hole act in opposite directions?
(b) Paragraph 10 – why is the final fracture dependent on P_2
    alone?
(c) Paragraph 29 – what is the meaning of the statement 'an
    appropriate conversion of peak baseplate force to axle
    load is to multiply by 4'? In your Fig. 6, peak
    baseplate loads of 200 kN are typical (for a 100 t tank
    wagon), implying that an axle load of 80 t can be
    expected.
(d) How are P_1 forces calculated?

    In paragraph 12 you mention that there is an impact force
due to differential wear at welds which will only be severe at
high speeds. However, the weld quality in alignment and
finish is relevant here, and high axle loads contribute
significantly to the impact force.

MR C. O. FREDERICK

Taking the queries in turn:

(a) The P_1 force occurs within 2–3 cm of the rail end, i.e. before the first bolt hole, whereas P_2 occurs well beyond the first bolt hole. Thus the direction of the shear force seen by the first bolt hole is reversed between these two events and the range of shear stress is related to the sum $P_1 + P_2$.

(b) The cracks which develop at the first bolt hole are most frequently inclined at $45°$ to the horizontal and lie in a plane such that fracture can detach a part of the rail end which includes the head of the rail. The final fracture is associated with the maximum tensile stresses which can develop across the crack tip. To develop a tensile stress across the crack tip requires a shear stress acting in a clockwise direction round the bolt hole for a train running from left to right. The maximum shear stress is generated by the P_2 force, thus final fracture depends on the crack size and P_2.

(c) Insofar as the axle load is equally divided between the rails and a wheel load is typically distributed on rail seats in the ratio 0.25:0.5:0.25 a factor of 4 is approximate in converting baseplate loads to axle loads. However, in Fig. 6 the force which is plotted is the peak value of a signal which represents the sum of four simultaneous baseplate loads. These baseplates were on the two sleepers immediately after the joint. At low speed, therefore, the maximum signal represents approximately 80% of the axle load; thus a 25 t axle load gives a maximum reading of 200 kN at low speed.

(d) The P_1 wheel force can be calculated according to the formula

$$P_1 = P_0 + 2\alpha V \left(\frac{K_h M_e}{1 + M_e / M_u} \right)^{1/2}$$

Where P_0 is the static wheel load, α is the ramp angle, V is the forward velocity of the vehicle and M_u is the unsprung mass per wheel. M_e is an effective rail mass and typically $M_e \approx 0.4M$ where M is half the track mass per metre. K_h is a linearized hertzian contact stiffness between the wheel and the rail. (Its value is dependent on P_1 and an iterative procedure is needed which converges rapidly.) At welded joints the rail profile does not form a simple ramp and so this formula is not suitable. To calculate the impact force in these conditions requires a consideration of the precise shape of the rail profile and can best be done on a computer which simulates the hertzian contact spring and the wheel and rail masses.

DR Y. SATO, Railway Technical Research Institute, Tokyo

With regard to the force acting between the wheel and the rail
due to the wheel flat, our measurements before the opening of
the Tokaido Shinkansen are shown in Figs. 1 and 2. It is
known that the maximum force and response appear at a low
speed of about 20 km/h for wheel flats [1, 2]. What is your
opinion about this, Mr Frederick?

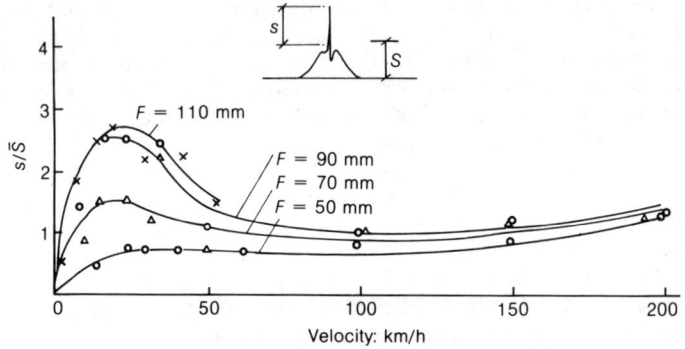

Fig. 1. Rail bending stress

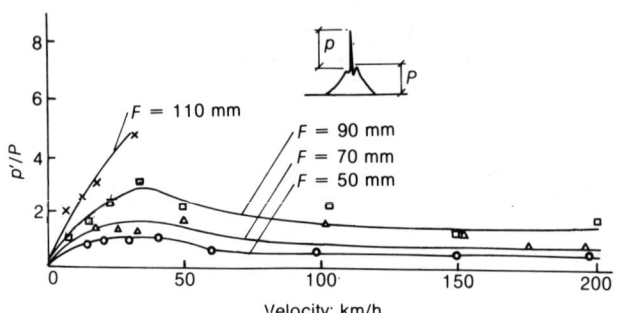

Fig. 2. Rail pressure

With regard to the control of roughness of the
circumferential wheel profiles, we have decreased the noise
level by 5 dB through the use of composite tread conditioning
blocks on the Shinkansen. The roughness was as shown in
Figs. 3 and 4 with tread conditioning blocks of composite and
cast iron [3].

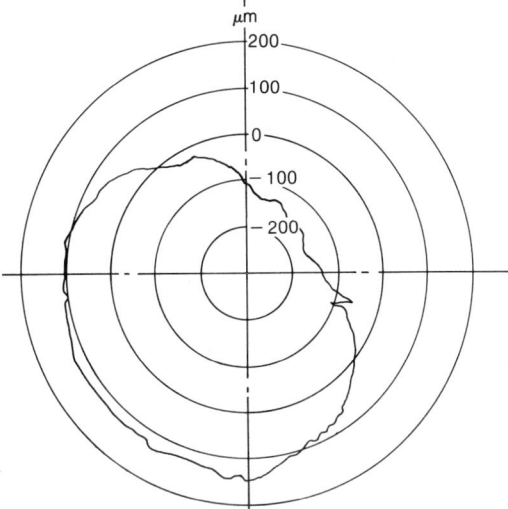

Fig. 3. Roughness of wheel tread with the use of a composite conditioning block

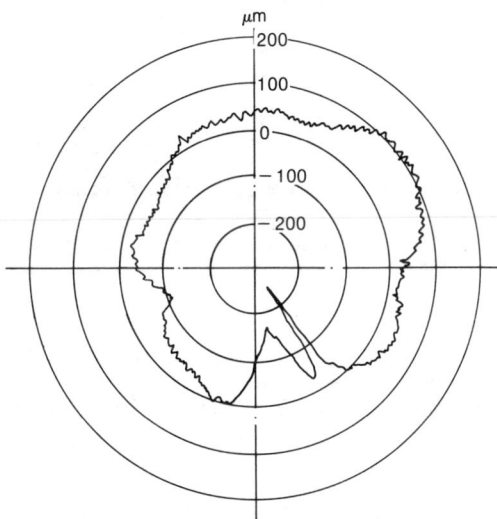

Fig. 4. Roughness of wheel tread with the use of a cast iron conditioning block

REFERENCES
1. Effect of flat wheels on track & equipment, Proc. Am. Rly Engng Ass., 53, 1952.
2. Betzhold, Ch. Erhöhung der Beanspruchung des Eisenbahnoberbaues durch Wechselwirkung zwischen Fahrzeug

und Oberbau, Glaser Annalen, 81, No. 3–5,Apr. 1957.
3.   Ohyama, T., Okamoto, I., and Chino, S. Influence of
       surface corrugation of wheel-tread on running noise in
       Shinkansen, Q. Rep. 19, No.4, 1978.

MR C. O. FREDERICK

When we measured the dynamic response of track to vehicles
traversing an indentation in the rail, we found that there was
a peak in the dynamic wheel-rail force at low speed and a
further peak at high speed. The latter peak was at P_1
frequency, i.e. 1000 Hz, so that its effect diminished with
increasing distance from the point of impact and the results
obtained will depend on where the strain is measured.
However, we know that wheel flat impact represents a serious
source of dynamic loading for concrete sleepers on high speed
lines. In general, the precise circumferential profile of the
wheel will decide at what speed the impact is most damaging.
    I am interested to learn that the Japanese National
Railways are using tread conditioning blocks and I assume this
is aimed at improving adhesion as well as smoothing the wheel
profile. I am not surprised that this reduces wheel noise.
We have found that disc-braked vehicles are noticeably quieter
than tread-braked vehicles, most probably because of less
wheel profile roughness. However, we believe that to obtain a
substantial improvement we must smooth both the wheel and the
rail. A corrugated rail will cause a great deal of noise even
with newly turned wheels.

DR M. H. MAGUED, Morrison Hershfield Ltd, Toronto

Firstly, with regard to the causes of track deterioration, we
are dealing with a complex phenomenon. Perhaps one cause of
track deterioration is the amount of energy that is to be
absorbed by the track structure and its subgrade. This is a
function of many parameters. This energy arises from the
interaction of the vehicle and the track, and this is a
function of the subgrade stiffness, gradation and strength,
the ballast characteristics and thickness, the tie size, mass
and thickness, the pad stiffness, the rail mass and section,
track irregularities, the wheel characteristics, the load and
speed, the track behaviour, the suspension stiffness and
damping, as well as the vehicle size and mass and the height
of the centre of gravity. The track design and upgrading must
be treated as a system. We cannot look for a single culprit
in any one element.
    Secondly, with regard to the use of the P_1 and P_2 loads
reported by Mr Frederick in the design of track elements, we
need to be aware that these are very fast phenomena and,
unless we know the behaviour of the element under high strain
rates, the use of the reported values in the design are not

appropriate. We do not have enough knowledge of the flexure
and bond strength of concrete ties under such high strain
rates. Similarly we do not know enough of the strain
behaviour of other components of the track structure and its
subgrade.

MR C. O. FREDERICK

Although I agree with Dr Magued that the track must be treated
as an overall system, and we have sought to do this at Derby,
nevertheless, in practical terms we must ask which parts of
that system are most significant and where can changes be made
to give the optimum economic advantage? We have also tried to
provide these answers.

On the subject of frequency behaviour of components, on
British Rail we are aware that P_1 forces especially are of
high frequency and that components do not act as rigid beams
etc. at these frequencies. The behaviour of rails and
sleepers can be calculated and values of damping inferred.
However, the rail pad behaviour is not well understood as its
characteristics change with both amplitude and rate of
loading. Several workers report studies to measure the
appropriate parameter values, but no one has convincingly
demonstrated the appropriateness of the measured values.

DR S. L. GRASSIE, Research & Development Engineer, Pandrol
Ltd, London

Our attention has been drawn to the damage done to track by
dynamic forces at what are, in conventional railway
engineering terms, relatively high frequencies. My
contribution refers to dynamic excitation at these frequencies
and what can be done to minimize damage.

It can be inferred from Fig. 7 of Paper 12 that the
relative displacement between the rail and the rail seat,
which is the deflection of the rail pad and the clip, at
frequencies of several hundred hertz is almost equal to the
amplitude of the imposed irregularity. Thus the lower the
impedance of the pad (i.e. the lower its stiffness and
damping), the smaller is the force transmitted by it to the
sleeper and the lower are the sleeper strains. In general, a
more resilient pad also gives lower contact forces at the rail
head: this is desirable for minimizing surface damage of the
rails.

It should also be borne in mind that sleepers used both
in metro lines and on conventional track are flexible beams
with little inherent damping. If damping from other sources
is also negligible, the concrete beams will vibrate with large
amplitudes under dynamic loading at the rail head, and
consequently also with large strains.

BIBLIOGRAPHY
Grassie, S. L., Gregory, R. W., Harrison, D. and Johnson, K.
L. L. (1982). The dynamic response of railway track to high
frequency vertical excitation. J. Mech. Engng Sci. 24, 77–90.
Grassie, S. L. and Cox, S. J. (1984). The dynamic response of
railway track with flexible sleepers to high frequency
vertical excitation. Proc. Instn Mech. Engrs 117–124.

MR H. M. ALEXANDER, Project Engineer, Western Australia
Government Railways, Perth

I am disappointed that what has been said about rail joints
refers only to the stresses they engender without offering a
means of reducing, minimizing or hopefully eliminating them.

I therefore put forward a suggestion on how this may be
achieved.

The present suspended joint is virtually the same as at
the beginning of the century and continues to have the
following weaknesses.

(a) It develops only 18–25% of the strength of the present
     rail.
(b) It dips causing stresses in the rail as high as twice the
     nominal stress.
(c) It causes differential settlement of both the ballast and
     the subgrade.
(d) It requires uneven sleeper spacing whereas modern
     mechanical track maintenance calls for a standard or
     uniform sleeper spacing.
(e) It creates uneven track response to train loading.

Although I do not think that any mechanical joint will
completely overcome these weaknesses, I suggest that a
supported joint based on the following principles (Fig. 1)
will help to minimize them.

(a) Cut the rail ends at an angle of between $66°$ and $72°$ to
     the longitudinal axis of the rail, leaving a gap of
     1/4 in or 3/8 in if the rail drilling pattern allows.

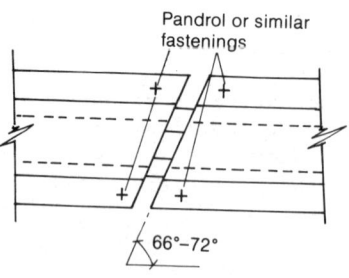

Fig. 1

(b) Use a thin section fishplate which will allow
standard fastenings, Pandrol or similar, to secure the
rail ends at four points as shown.

The object is to provide a joint which allows a wheel
tyre to pass over it in such a way that the wheel-rail contact
is maintained as the wheel passes. This will ensure minimum
line loading and vibration and therefore minimum differential
ballast or formation settlement, while simultaneously allowing
a uniform sleeper spacing through the joints, ensuring a near-
uniform track response to loading.

Since track joints nowadays are mostly insulated joints,
which are even weaker than normal joints, the supported joint
has much to offer.

Mr Frederick, do you think that the vertical loading
advantages claimed would be achieved, assuming that in
concrete-sleepered track a timber sleeper of the same depth as
the concrete sleepers would be used to support the joint?

MR C. O. FREDERICK

In general I agree with you that it would be possible to
produce a stronger design of joint. The restraints against
doing so are mainly of a practical nature but a supported
joint should be better than an unsupported joint. However,
the current problem is how to deal with the population of
joints we already have and if they are already crippled it
would not help to change their support conditions since we
would increase the impact force owing to the greater effective
track mass. For the future the welded joint should be our
target not a bolted joint of any kind. For insulated joints
in long welded rail we currently use glued insulated joints
and we are seeking to strengthen these joints. Your proposed
design may help to reduce the rail end impact; alternatively
it may be found that the projecting 'hose' of metal is more
prone to plastic deformation or fracture. The use of timber
sleepers to provide more resilience and less mass and thereby
to reduce further the impact forces could help to improve the
life.

DR C. ESVELD, Head of Railway Technology and Quality Control,
Nederlandse Spoorwegen, Utrecht

Mr Frederick stated that the rail (including the welds) should
be made as smooth as possible, a specification which cannot be
achieved in practice. The examples shown dealt with weld dips
between 1 mrad and 10 mrad. The limiting value should be
1.2 mrad, corresponding to 0.3 mm on a 1 m base. From
experience on the Nederlandse Spoorwegen beyond 2 mrad
problems increase rapidly. Cracking of concrete sleepers
occurred beyond 4 mrad. The solution is to straighten the
welds. This has been achieved with the new STRAIT concept in

167

combination with grinding by the Plasser GWM 220, at present used by Nederlandse Spoorwegen, both in the welding depot and as mobile machines for in-track operation. With this approach the quality achieved is much better than 1.2 mrad. Consequently most of the problems on weld geometry deviations have now been solved.

MR C. O. FREDERICK

I am also an advocate of straightening and smoothing rail welds in track and agree with the tolerances you suggest.

MR M. J. SMITH, Area Civil Engineer, British Rail, Norwich

I would like to ask Mr Frederick two questions related to joints which are at present having to be retained in our high speed, mixed traffic, continuously welded rail track.

On electrified routes there is still a need for insulated joints. Although improvements to their working performance have been made, it is my experience that these joints will have to be replaced at least once before the rail is cascaded. From your research into P_1 and P_2 forces on joints, do you have any suggestions which may curtail premature failure, e.g. should the first bolt hole have a smaller diameter, or should the plate be lengthened to ensure that it becomes semi-supported, or would a six-hole plate be better?

With regard to paragraph 11, are your observations analogous to the difficulty in maintaining tight joints near cast manganese crossings? I believe that cast crossings have a stiffness that is unlike that of the adjoining plane rail and that joints, like welds, should be 3 m from such a change in stiffness. Your research indicates a need to review the casting to bring about a more gradual change in stiffness or a crossing unit which is less stiff. What are your views on this?

MR C. O. FREDERICK

The behaviour of glued insulated rail joints is surprisingly complex and is not yet completely understood. Most failures seem to involve fractures of bolts or bolt holes, but it is suspected that these are secondary effects resulting from an earlier failure of the adhesive bonding. One hypothesis is that glues which are flexible enough to avoid fracture due to impact are also characterized by a tendency to relax under load. Thus, the initial bolt tensions are reduced, so that the glue strains are increased and failure is induced.

To lengthen the plate, as suggested, would reduce the risk of bolt or bolt hole failure. Similarly, to deepen the plate could result in a joint with more uniform flexibility and hence smaller dynamic strains. From a practical point of view, either of these changes would increase weight and expense and this would have to be balanced against the inconvenience of joint replacements.

Studies of this problem are in hand at Derby and trials of improved joints are expected to be in progress before the end of 1985.

I agree that the change in bending stiffness at the tight joint between a cast crossing and the abutting plain rail must worsen the stress environment of these joints. In effect, the joint will be obliged to carry higher bending moments owing to the rigidity of support it receives from the crossing. However, there are other factors which make these joints a problem.

There is the difficulty in obtaining a good fit between wrought and cast components and there is the mismatch in strength properties between BS11 normal quality steel and AMS which creates step and impact loads. The problem can be eased by shortening and lightening the casting so that the casting does not provide so much rigidity. It can also be eased by using a bainitic alloy for the casting so that the legs can be welded on. If bolted, joints must be used; they can then be placed wherever it is most convenient.

13   Improved steels and metallurgy

J. D. YOUNG, Technical Co-ordination Manager, British Steel Corporation, Workington

SYNOPSIS. The increasing demands on track are reflected in higher rail steel quality requirements. Further development of the current range of rail steels, already based on fully pearlitic microstructures is limited by the need to avoid hyper-eutectoid cementite. However narrow chemistry ranges; better chemical and mechanical uniformity, and cleaner steel would be beneficial. Tighter geometric requirements are outlined. The rail manufacturing improvements achieved recently and their probable further development are discussed with reference also to potential constraints within metallurgical and manufacturing technology and also within acceptable costs.

INTRODUCTION

1. Traditional railway design and operating philosophy, based on steel wheels running on steel rails, is not expected to change in the next decade. Likewise it is expected that trends toward higher traffic speeds; faster acceleration and more efficient braking; heavier axle loads; increased track utilisation; higher standards of passenger comfort; and demands for reduced track maintenance costs will continue to intensify.

2. These railway operating developments generally expose steel rails to increasingly severe service conditions and stress environments. Rails must, therefore, be produced to the highest possible quality standards, albeit at relatively low product cost.

3. The metallurgy of rail steels and the geometric/engineering properties of the rails are, and will continue to be, the subject of study by both users and makers.

RAIL METALLURGY

4. The overall objectives of long life and low service failure risk involve various metallurgical and mechanical properties including chemical composition; tensile and yield strength and hardness; ductility and toughness; fatigue and crack growth characteristics; and residual stress levels in the finished new rails. The optimum combination of these

properties must be chosen for a particular track design and its operating conditions.

5. The evolutionary inter-action between rail users and makers, combined with strong economic forces, had led to carbon-manganese steels of fully (or almost) pearlitic microstructure being almost universally adopted for rails. Whilst small tonnages of steels of other microstructures eg ferritic for electrical conductor rails, and austenitic high manganese for crossing rails, are currently used; and experiments with special composition crossing rails with bainitic structures are in process, it is anticipated that pearlitic microstructures will remain the preferred choice for the foreseeable future.

6. Pearlitic rail steels can be roughly classified into three major groups:-
Normal grade of minimum U.T.S. = 680 N/sq.mm. eg British Standard (BS) 11, Normal, and
International Union of Railways (U.I.C.) 860, Grade 70;
Wear Resisting grade of minimum U.T.S. = 880 N/sq.mm. eg BS11, Wear Resisting Grades A and B, UIC 860, Grades 90 A and 90 B and American Railway Engineering Association (AREA) Standard Rail;
High Strength (or Premium) Grade of typical UTS = 1080N/sq.mm. eg AREA High Strength Rail. The two main types include alloy rails (generally containing about 1% chromium and sometimes traces of other alloys); and heat treated wear resisting grade rails eg. BS11 Wear Resisting A Grade or AREA Standard carbon grade.

7. Being generally supplied in the as-rolled and cold finished conditions, pearlitic steel rails of all grades have low toughness properties and are very notch-sensitive. In practical experience, the higher notch sensitivity (lower toughness) of high strength grade and wear resisting rails compared with normal grades, is very apparent, necessitating extra precautions in manufacture, handling and usage. Stricter control of steel hydrogen content and cleanness and of surface quality is essential. The various toughness measurement (including fracture toughness) techniques generally place the high strength rail grades towards the bottom of the overall rail steel scatter band - which is low in comparison with many other steel products eg structural steel. The higher strength rails (with associated higher yield strengths) tend to have higher residual stress levels after roller straightening. Also some types require special precautions during rail welding operations.

8. Nevertheless monitored service performance tests in curved track prone to rail side cutting, confirm the superior performance of higher strength, harder grades as indicated by the following typical wear resistance ratios -

Normal Grade	1.0
Wear Resisting	1.7 - 2.0
High Strength	4.0 - 5.0

9. The next decade's more severe track service conditions

is likely to result in an increasing proportion of wear resisting and high strength rails being used, particularly for heavier axle loads and tightly curved track. However a continuing role for all three rail grade groups is expected. As with many other components, there are dangers in specifying on one property eg wear resistance, at the risk of fore-shortened life for other reasons eg surface fatigue defects.

RAIL STEEL MICROSTRUCTURE

10. The abrasive wear resistance and ability of rail steel to withstand plastic deformation and cold mechanical working in the highly stressed wheel contact zone is derived from its pearlitic structure.

11. Pearlite grains basically consist of alternate microscopic lamellae of hard iron carbide (cementite) and soft, ductile pure iron (ferrite). This composite microstructure has a combination of mechanical properties different from either of the constituent components. The size (on a micro scale) of the grains, and especially the thickness of the lamellae (or inter-lamellar spacing) have a marked influence on the tensile and yield strengths and associated hardness and wear resistance of rails.

12. Normal grade rail steels, by dint of milder chemistry, have a microstructure consisting of a grain boundary network of soft, ductile iron (ferrite) surrounding pearlite grains of relatively coarse lamellae. This produces lower tensile strength and hardness, less notch sensitivity/higher toughness and lower hydrogen sensitivity. The wear resisting grades of higher tensile strength/hardness; higher notch and hydrogen sensitivity have fully pearlitic microstructures with finer/thinner lamellae. The high strength grades are also fully pearlitic in microstructure, ith higher strength and hardness being achieved by further refinement of the pearlite lamallae through either chemical alloying (the addition of chromium and in some cases other microalloys) or by heat treatment.

However there is a metallurgical barrier or "ceiling" to further extrapolation. Beyond approximately 0.82% carbon, the microstructure becomes hyper-eutectoid containing an interlacing grain boundary network of hard, extremely brittle iron carbide (cementite), incapable of withstanding impact loading, and hence prone to lead to catastrophic multiple service fracture if present in rails. Also excessive refinement of a pearlitic structure can lead to break-up of the lamellar structure and "balling-up" or spheroidizing of the cementite resulting in considerable loss in hardness. Whilst minor adjustments in this metallurgical "ceiling" are possible, the main objective in high strength rail manufacture must be to work within the tightest possible scatter-band just below the ceiling.

13. Other technologies, eg rail head profiling and better more effective rail lubrication etc, have proved effective

in reducing rail wear and their use is likely to increase. This may reduce the demand for an increasing proportion of high strength rails and for increasing tensile strengths and hardness values. However increase of rail life and decreased rail wear will focus increasing attention on the fatigue and fracture behaviour of rails. Arising from such studies there is an obvious demand for more uniformity and consistency of internal and surface quality of rails.

14. Thus the predicted possible trends in railway operation are likely to increase the need for -
- (a) Tighter control of rail steel chemical composition
- (b) More uniformity of chemical composition within the rail cross-section, length and cast
- (c) minimal segregation and micro-segregation of chemical elements
- (d) Cleaner steel having a lower total content of non-metallics and absence of both large sized and angular shaped inclusions
- (e) Absence of piping, lamination and macro-segregation
- (f) Absence of hydrogen shatter-cracks (flakes)
- (g) Excellent rail surface quality - with absence of "notch" type defects
- (h) Minimal surface decarburisation

RAIL GEOMETRY

15. The current and predicted future railway operating conditions generate increasing demands for more accurate track geometry and alignment, for smoother and more accurately shaped rail head surfaces, and better control of rail weld geometry and metallurgical structure. Thus more stringent geometrical requirements must be achieved during rail manufacture, including:-
- (a) Accurate section dimensions with tight tolerances
- (b) Accurate verticality and symmetry
- (c) Tighter general, and particularly end straightness standards
- (d) Avoidance of both short and long wavelength undulations on the rail head crown
- (e) Improved control of rail head crown radius profile
- (f) A higher demand for rails of uniformly long lengths
- (g) Minimal residual stresses

RAIL MANUFACTURE

16. The previous paragraphs are based on practical track experience and more theoretical metallurgical and engineering studies and indicate the various metallugical and geometric improvements desirable in rails for the future. The following paragraphs outline the advances made in rail manufacture in the past 10-15 years and ongoing developments and investigations in the railmaking industry to meet the requirements of the next decade.

17. Steelmaking. The older traditional steelmaking processes eg Acid Bessemer, Thomas and Basic Open Hearth are

virtually obsolete. Most rail steel is now and in future is likely to be manufactured by the Basic Oxygen Steelmaking process, or variants thereof. This highly productive process achieves more consistent and predictable chemical reactions (and hence chemical composition of the product) than the older processes and is thus capable of instrumentation and computer control. Residual element and gaseous element contents are more uniform and steel cleanness is improved. A minority tonnage of rail steel is produced by the oxygen assisted electric arc process, generally under improved metallurgical control as compared with the earlier electric arc processes.

18. Various types of secondary steelmaking processes are likely to be more widely adopted for improved liquid rail steel refinement in the future, in order to achieve more accurate control of chemical composition, temperature, hydrogen, steel deoxidisation and cleanness. These secondary steelmaking processes include vacuum degassing, inert gas stirring, chemical composition, trimming, ladle slag control and additive injection techniques.

19. Liquid rail steel casting. The long established process of casting liquid steel by top pouring into open topped ingot moulds as semi-killed (or balanced) ingots was prone to many internal, sub-surface and surface defects, and is now virtually obsolete.

20. In 1974, British Steel Corporation were the first railmaker to adopt the modern process of continuous casting of blooms for its entire rail production. The major and highly significant improvements in rail quality achieved include elimination of central piping, lamination and associated macro-segregation, a major improvement in rail surface quality, cleaner rails with lower total inclusion content and smaller sized inclusions, and improved uniformity of chemical composition and mechanical properties.

21. Most other railmakers of international repute have now adopted the continuous casting of rail steel and this process is likely to become the norm in the next decade.

22. Refinements in continuous casting of rail steel which are in various stages of development, include total enclosure of the liquid steel during casting, using special refractory tubes and inert gas shrouding; elecromagnetic stirring of the liquid steel core of the bloom during its solidification; surface lubricant improvment; better control of heat extraction and hence steel solidification. These techniques have the objectives of further improving the bloom and hence resultant rail quality by improving steel cleanness, achieving even more uniformity of chemical composition and less segregation, and reduction of internal and surface imperfections.

23. Hot rolling of rails. The use of continuously cast blooms for rail manufacture has involved the installation in rail rolling mills of modern fully instrumented and

175

automatically controlled reheating furnaces. These new installations achieve more accurate and uniform bloom temperature and furnace atmospheres, thereby assisting in the improved control of rail section dimensions, rail surface decarburisation and metallurgical microstructure.

24. Progress made in recent years in improving roll pass designs and roll materials, rolling mill bearings, roll cooling and lubrication will undoubtedly be subject to further investigation and development; the major objective being the improvement of rail quality to meet the requirements of the future.

25. The hot rolling of rails incurs the use of heavy plant and equipment operating under arduous conditions of elevated temperatures and high loading. Hence there must eventually be limits to the degree of geometric accuracy achievable at an economic cost.

26. Most railmakers have made considerable investments in the cooling beds over which the as-rolled rails are progressed en route to finishing operations - to achieve better uniformity of as cooled rail shape (straightness) and reduced risk of mechanical damage to the rails.

27. Some makers have also installed equipment for acceleration of the air cooling to achieve finer pearlitic structures, and associated higher hardness and tensile strength, in the vitally important rail head wear zone.

28. Other investigations in process include the accelerated water cooling of either the full rail section or the head only to achieve strength and hardness levels of the high strength rail grade, utilising the residual heat content of the rail immediately after rolling. This concept is not new, and the service performance of the rails treated many years ago by a similar technique were not satisfactory. However the improved rail steel quality available currently and the availability of modern instrumentation and computer control may lead to the future evolution of heat treated rails of superior wear resistance than currently available high strength rails.

29. Rail Length. The welded joint is still a significant weakness in the track. Longer rail lengths of up to 36 m can now be delivered from modern mills.

30. Rail finishing operations. Roller straightening is currently the only economic way of achieving the desired high standards of rail straightness at an economic cost. The possible presence of either long or short wavelength undulations on rail head crowns and the residual stress levels present in roller straightened rails (and particularly in high strength grade rails), is cause for concern by some railway engineers.

31. By strict attention to detail in the maintenance, setting, adjustment and operation of rail roller straightening machines the head crown undulation problem can be kept within reasonable limits. However this is another example of the maker being compelled to operate near the

limits or "ceiling" of existing technology.

32. Trials are now in process with rails which have been "stretch straightened" using the principle long established on other lighter section steel products. Initial results indicate that such rails have low residual stress levels, but the achievement of the desired high standards of rail straightness from pilot plant or full scale production equipment, at economic costs, have yet to be proven. Also it will be necessary to prove that the track performance of stretch straightened rails is satisfactory and that there are no undesirable secondary effects arising from service induced residual stress patterns.

33. Rail inspection and quality assurance. The use of continuous, automatic, in line ultrasonic testing equipment for ensuring the absence of internal flaws/defects in rails, is a well established technique. Further refinement of these techniques and equipment is likely in future.

34. Traditionally other rail inspection procedures eg straightness checking and surface defect detection have been visual and manual operations and section dimensional measurement a manual process confined to a single "sampling" location.

35. Automated instrumental techniques developed for the inspection of other simple shaped steel products are being further extended for potential application to rail inspection and quality control, and some experimental or pilot plant installations are in operation.

36. It is expected that during the next decade, rail inspection and acceptance procedures will depend on the use of continuous automatic equipment installed in the production line for the checking, measurement, assessment and recording of rail section dimensions, general straightness, end straightness, head crown undulations, and surface defects.

CONCLUSIONS

37. Rail steels of various grades are currently available to meet the varied and increasingly demanding service conditions of today and the future. Many of the metallurgical and geometrical improvements desired can be identified and the many fundamental changes made in the rail manufacturing process route in the past 10 years have contributed towards these desired high quality standards. However existing metallurgical and manufacturing technology is being stretched towards its limits, and also by the constraints of acceptable and economic rail costs. Despite such difficulties, considerable effort will continue in the next decade to further improve rail quality and metallurgy.

14 Rail wear/fatigue limits

S. MARICH, BSc, PhD, Senior Principal Officer, BHP Melbourne Research
Laboratories, Broken Hill Proprietary Co Ltd, Australia

SYNOPSIS.   Historically, rail wear limits have been
related to practical experience and safety aspects of train
operations.   However, the recent introduction of new
technologies has reduced the importance of these criteria.
This paper describes the work conducted both in the
laboratory and in the field to quantify the stress
environment imposed on rails.   With these data, fracture
mechanics   principles   together   with   the   fatigue
characteristics of the materials are used to determine the
"safe" wear limits of rails subjected to specific operating
conditions.

INTRODUCTION

    1. The replacement of rails is a major source of
operating expenditure and potential loss of revenue
incurred by high axle load, unit train operations.

    2. Rail wear, plastic deformation and fatigue are the
main mechanisms controlling rail life.   Both wear and
deformation being longer term cumulative processes, can be
readily monitored by track maintenance personnel and hence
suitably incorporated in rail replacement strategies.   Rail
fatigue, however, is a major cause of concern because: (a)
fatigue defects on reaching critical size can lead to
complete fracture of the rail section and hence become
potential sites for derailments;   (b) the detection of such
defects relies on frequent and expensive ultrasonic
inspection of the track;   (c) replacements of defective
rails leads to considerable maintenance costs and
interference to operating schedules, and (d) the amount of
wear which a rail can tolerate is directly influenced by
the fatigue characteristics of the material.   Furthermore,
in recent years, the relative importance of fatigue as a
rail life controlling mechanism has increased with the
introduction of high strength rails, improved lubrication
practices   and   modified   wheel/rail   interaction
characteristics, all of which have led to a marked
reduction in the wear and deformation of rails,
particularly in curved track (refs. 1 to 3).

3. In any structural material, and rails are no exception, it is often not technically nor economically possible to prevent the occurrence of defects or small cracks. It is therefore essential to have a means of assessing their.behaviour and effect. Fracture mechanics provides an accepted basis for quantifying the characteristics of cracks in structures during their initiation, growth and final failure phases. The following sections give details on how fracture mechanics principles can be used together with the fatigue properties of the material to provide guidelines for realistic manufacturing and design purposes. Particular emphasis is placed on the initiation and early crack growth stages of fatigue which may develop from the surface of the rail head, although similar principles can be applied to other locations on the rail surface.

FATIGUE FROM SURFACE DEFECTS
General Concepts
    1. The initiation of fatigue cracks from surface imperfections in rails has not received detailed attention. The first stage of the current study, therefore, aimed to characterise the initial stages of fatigue crack growth and assess the validity of established fracture mechanics analyses.
    2. The basic equation relating the depth (a) of a <u>sharp</u> surface crack to the applied stress range ($\Delta\sigma = \sigma_{max} - \sigma_{min}$) is:

$$\Delta K = Q \, \Delta\sigma \, \sqrt{\pi a} \tag{1}$$

where ΔK = stress intensity factor range, and
           Q = constant which depends on the geometries of the crack and the component.

    3. Since this study concentrates on the nucleation of fatigue cracks, use can be made of the above equation rewritten in the form:

$$a_{th} = \frac{1}{\pi} \left(\frac{\Delta K_{th}}{Q \Delta\sigma} \right)^2 \tag{2}$$

where a_{th} is the maximum surface crack depth which can be tolerated without causing fatigue crack growth and ΔK_{th} is the threshold stress intensity fracture range (a measurable material parameter).
    4. The procedures for determining the parameter Q have been summarised in ref. 4 for a wide variety of crack geometries. For semi-elliptical surface cracks in a slab subjected to a uniform uniaxial tensile stress, at a crack depth to slab thickness ratio of less than 0.1, and for cracks of depth to half width (c) ratios of 0.4 and 0.5, the values of Q are 0.94 and 0.88 respectively. Therefore equation (2) becomes:

180

$$a_{th} = 0.36 \frac{\Delta K_{th}^{2}}{\Delta \sigma} \quad \text{for } a/c = 0.4 \quad (3a)$$

and

$$a_{th} = 0.41 \frac{\Delta K_{th}^{2}}{\Delta \sigma} \quad \text{for } a/c = 0.5 \quad (3b)$$

5. Fig. 1 shows the fatigue crack growth characteristics of A.R.E.A. standard carbon rails, determined in a previous study using compact tension specimens. It can be seen that the measured value of ΔK_{th} is in the range 12 to 15 MPa√m. With this information acting as a guideline, equation (3) can be used to obtain the relationship between a_{th} and $\Delta \sigma$ for various values of ΔK_{th} (the values chosen are at the lower end of the measured range), as shown in Fig. 2. Thus, if the above concept is valid, at a particular $\Delta \sigma$ value cracks with depths below the curves are not expected to exhibit growth.

Experimental Procedure

1. The fracture mechanics approach is based on the analysis of sharp cracks. The methodology therefore is very conservative in nature since the majority of surface defects which may be introduced during rail manufacture, such as scale and handling marks, are relatively blunt. Nevertheless, to be consistent with the approach, sharp (0.06mm root radius) notches of varying depths were machined by means of spark cutting, at several locations on the head of the rail specimens, as shown schematically in Fig. 3. Copper discs of various diameters were used for the operation such that for a particular crack depth, the resultant a/c ratio was in the range 0.4 to 0.5. Along the rail length, the notches were no less than 50mm apart to limit the influence of one notch on the stress intensity factor at the tip of the adjoining notch to less than 2%. The spark cutting operation also produced a thin layer of martensite on the surface of the notch which enhanced the severity of the test procedure.

2. The notched rail samples were subjected to fatigue loading conditions using a rolling load machine which has been described in detail previously (ref. 5). The samples were assembled in the machine on a baseplate canted at 6° from the vertical (Fig. 3) to simulate uneven loading of the rail section and hence any torsional affects. All tests were conducted at a vertical applied load of 150 kN, equivalent to a nominal axle load of 30 tonnes. As illustrated in Fig. 3, the wheel segment travelled a distance of 300mm over the rail running surface, subjecting both the supported and cantilevered sections of the rail to loading conditions. The rate of loading was four complete wheel cycles (back and forth) per second.

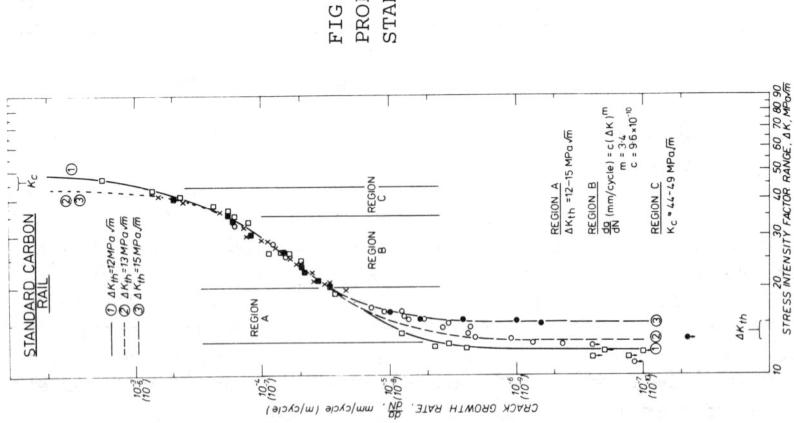

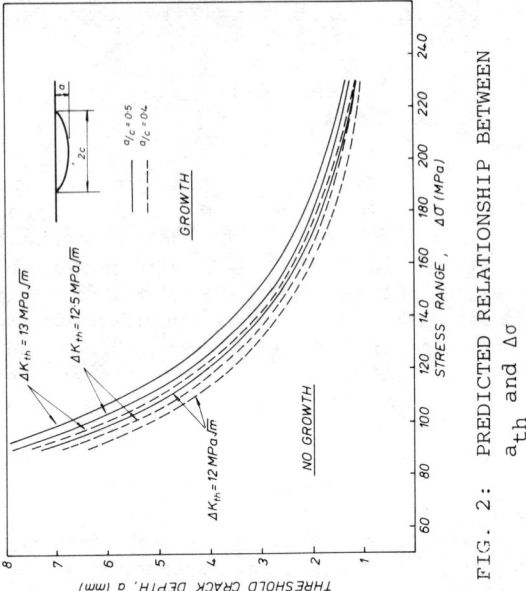

FIG. 2: PREDICTED RELATIONSHIP BETWEEN a_{th} and $\Delta\sigma$

FIG. 1: FATIGUE
PROPERTIES OF
STANDARD RAIL.

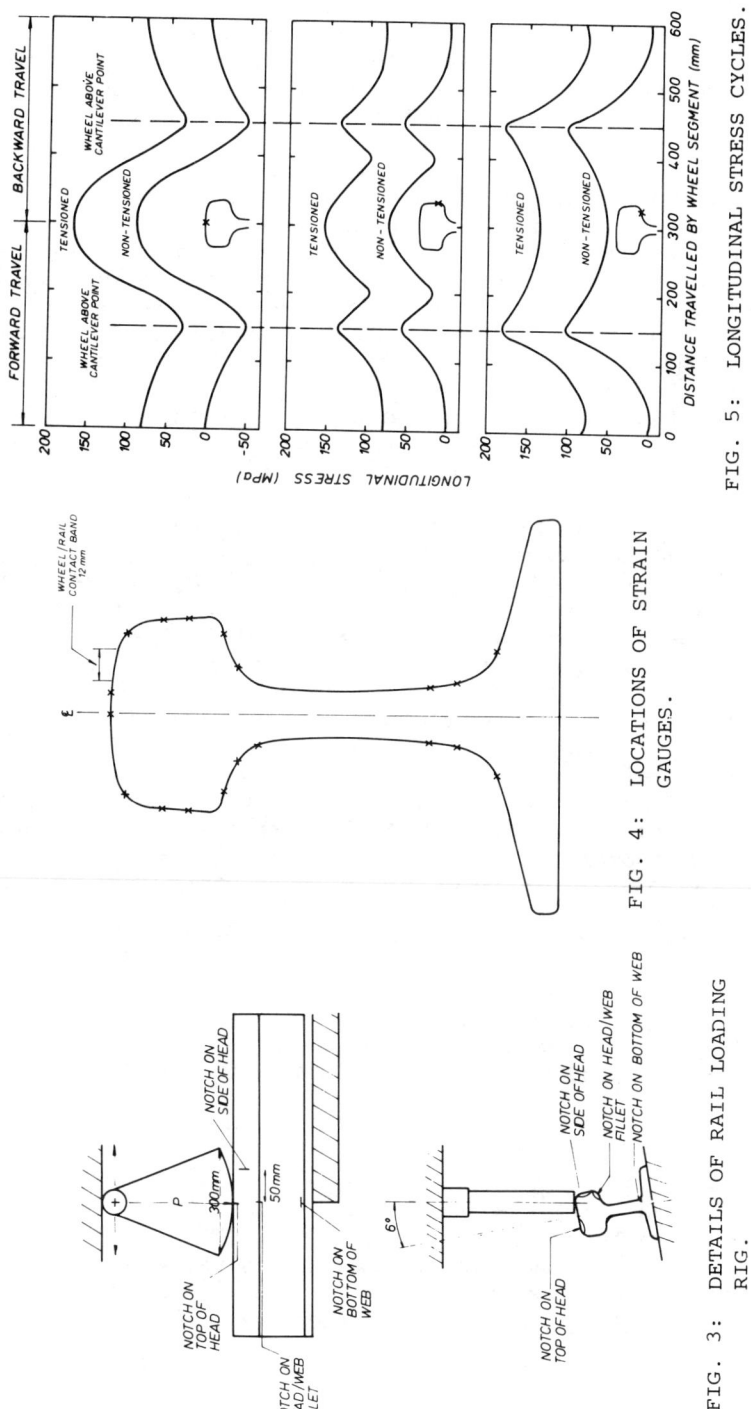

FIG. 5 : LONGITUDINAL STRESS CYCLES.

FIG. 4 : LOCATIONS OF STRAIN GAUGES.

FIG. 3 : DETAILS OF RAIL LOADING RIG.

3. Some of the specimens were also fitted with a rail tensioning rig, designed to apply constant longitudinal tensile stresses equivalent to those generated by thermal contraction of the rail at temperatures below the stress free temperature of rail welds. In this study, the rails were subjected to a stress of 80 MPa, i.e. equivalent to 35°C below the stress free temperature.

4. To quantify the mechanical stresses applied to the rails during fatigue loading, strain gauge rosettes were applied at numerous locations on the rail cross section, as shown schematically in Fig. 4, and at three positions along the length of the rail, namely at the cantilever point and at 50mm on either side of this point. Strain readings from the gauges were recorded as the rolling load travelled over a complete cycle.

5. Examples of the variation in the longitudinal stress measured at three locations on the rail cross section at the cantilever point are shown in Fig. 5. It is of interest to note that on the top and side of the rail head the maximum tensile stress is obtained when the wheel segment is at the end of its travel on the unsupported rail section, i.e. maximum bending. At the bottom of the rail head, on the other hand, the maximum tensile stress occurs when the wheel segment is directly above the measuring point, i.e. the resultant stress is due to the bending of the rail head on the web. Because of this bending, compressive longitudinal stresses are also obtained at the top of the rail head. Pretensioning of the rail results in an overall increase in the tensile stress.

6. In service, rails are subjected to a total longitudinal stress (σ_T) which arises from several sources, namely:

.   thermal stresses due to cyclic temperature variation, σ_t,
.   general bending stresses due to lateral and vertical wheel loads, σ_b,
.   residual stresses due to manufacturing procedures and in-service conditions, σ_r, and
.   localised contact stresses due to wheel/rail interaction, σ_c.

Therefore,

$$\sigma_T = \sigma_t + \sigma_b + \sigma_r + \sigma_c \tag{4}$$

7. The parameter $\Delta\sigma$ in equation (3) must take into account all of the above stresses. As mentioned previously, in the current study, bending and simulated thermal stresses were measured by strain gauging the rail section. Residual stresses, which may vary considerably depending on the manufacturing procedure, were also measured using established strain gauging and sectioning techniques. Fig. 6 illustrates the results obtained. Contact stresses, on the other hand, were ignored in the

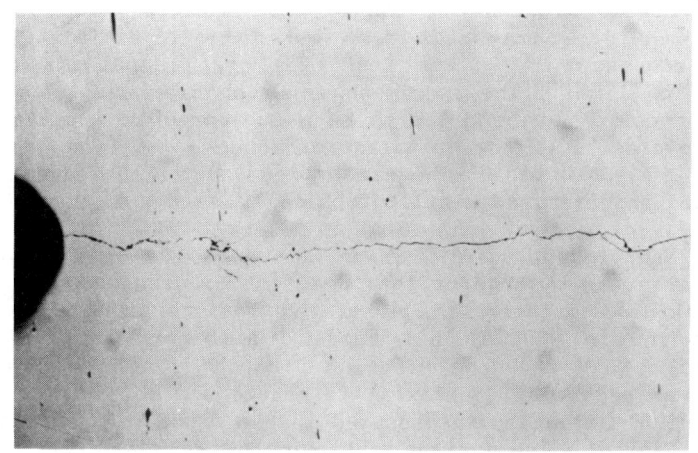

FIG. 7: CRACK GROWTH FROM NOTCH TIP.

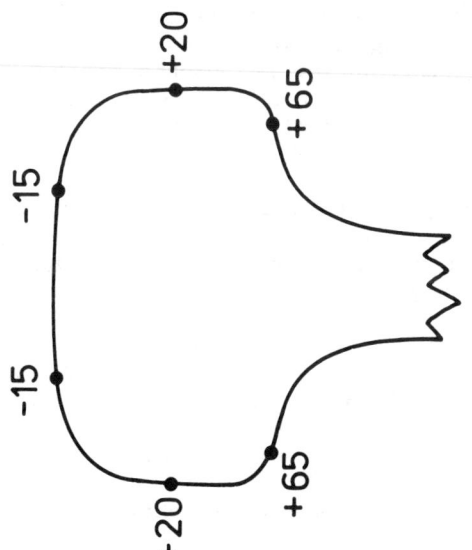

FIG. 6: LONGITUDINAL RESIDUAL STRESSES IN AS-ROLLED RAIL.

analysis since the localised effect of wheel/rail interaction was maintained at a considerable distance from the surface notches.

8. Crack growth, if any, was measured using a Krautkramer d.c. potential drop meter which exhibited a proven accuracy of $\pm$ 0.2mm for total crack depths in the range 2 to 15mm. The specimens in which crack growth was small or could not be detected were subjected to 1 million stress cycles, i.e. an equivalent of about 40 million gross tonnes of traffic. These samples were subsequently sectioned and examined metallographically.

Results and Discussion

1. The early stage of fatigue crack growth from one of the machined notches is illustrated by the longitudinal section in Fig. 7. In all tests, the crack propagation occurred on a transverse plane, indicating that the major stress was essentially uniaxial in nature and acting in the longitudinal direction, as assumed in the fracture analysis.

2. Figure 8 summarises the results obtained in the form of crack growth/no-growth as a function of initial crack depth and measured stress range (including bending, thermal and residual stresses). Each point on the graph represents an individual initial machined notch. The curves relating threshold crack depth to stress range obtained from equation (3) for a ΔK_{th} value of 12.5 MPa $\sqrt{m}$ and for a/c values of 0.4 and 0.5 are also shown in the figure. It is evident that the results validate the fracture mechanics concept, i.e. at a particular $\Delta\sigma$ value cracks with depths below the curves <u>do not</u> exhibit growth.

3. In achieving the validation, two aspects related to the results require comment:

(a) As shown in Fig. 5, notches on the side and bottom of the rail head in non-tensioned rails were subjected to a stress cycle with $R \sim 0$ ($R = \sigma_{max}/\sigma_{min}$). The respective $\Delta\sigma$ values indicated in Fig. 8 were therefore equivalent to σ_{max}.

(b) Notches in tensioned rails were subjected to positive R values. For these, the stress values plotted in Fig. 8 were also equivalent to σ_{max} (and not $\Delta\sigma$). This may be justified as follows:

Cooke and Beevers (refs. 6 and 7) have shown that in rail steels the relationship between ΔK_{th} and positive R values can be approximated by the equation

$$\Delta K_{thR} = \Delta K_{th} (1 - R) \tag{5}$$

where ΔK_{thR} is the threshold at a given R ratio and ΔK_{th} is the threshold for R = 0, i.e. the value obtained experimentally and used in this study.

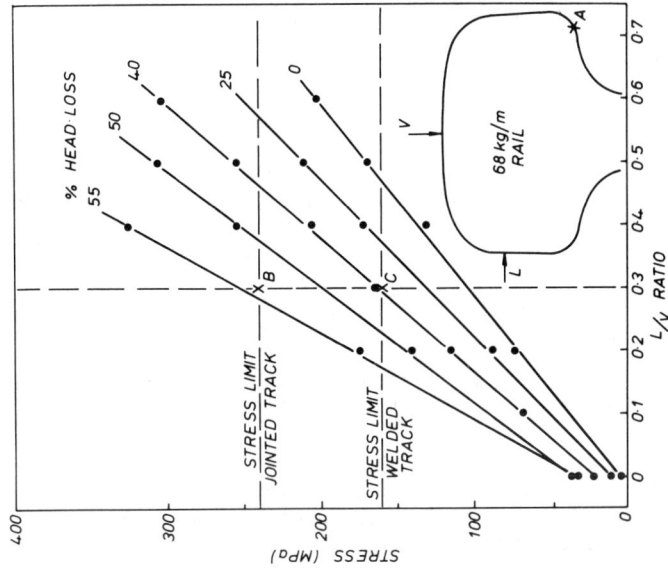

FIG. 9: INFLUENCE OF RAIL HEAD LOSS ON STRESS.

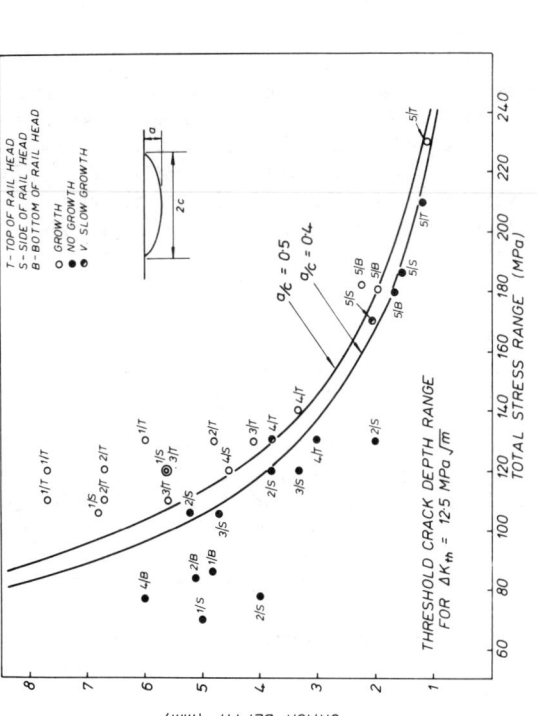

FIG. 8: TEST RESULTS.

Equation (2) therefore can be rewritten as:

$$a_{thR} = \frac{1}{\pi} \frac{\Delta K_{th} (1 - R)}{Q \Delta \sigma}^{2} \tag{6}$$

where a_{thR} is the threshold crack depth at a given R ratio.

Since $R = \sigma_{min}/\sigma_{max}$ and $\Delta \sigma = \sigma_{max} - \sigma_{min}$, equation (6) becomes

$$a_{thR} = \frac{1}{\pi} \frac{\Delta K_{th}}{Q \sigma_{max}}^{2} \tag{7}$$

The use of σ_{max} values in plotting the data, therefore, accounts for the influence of R on a_{th}.

RAIL WEAR LIMITS

1. The deferment of rail replacement by, for example, extending the acceptable rail wear limits can lead to substantial cost savings.

2. Laboratory and field experiments have shown that the maximum longitudinal tensile stresses due to bending and torsion of a rail are obtained at the field lower corner of the rail head (point A in Fig. 9). The graphs in Fig. 9 show measured stresses at various lateral to vertical (L/V) force ratios (V = 150 kN) and percent rail head losses, for a 68 kg/m rail section and central rail loading. From Fig. 8 it can be seen that a 1mm deep sharp notch will not exhibit fatigue growth at a stress level of 240 MPa. Thus, for jointed railway track, in which thermal stresses are not significant, the data in Fig. 9 indicate that at a L/V ratio of 0.3 (typical of 600 to 800m radius curves) a rail head loss limit of just under 55% (point B) would be acceptable in terms of fatigue performance. In welded rail, longitudinal tensile thermal stresses have to be considered. For a thermal stress of 80 MPa (equivalent to a rail temperature of 35°C below the stress free temperature), the allowable head wear limit is 40% (point C). Both of these values are considerably greater than what is currently accepted.

3. A similar procedure may be used to assess the potential influence of other operating variables such as axle loads, vehicle speed and track geometry at any required location on the rail surface. It must be emphasised however, that the analysis does require an accurate knowledge of the total stresses imposed on the rails. These may vary considerably from system to system depending on the specific wheel/rail interaction characteristics and in particular the location of wheel contact on the rail's running surface.

CONCLUSIONS

1. For rail steels the nucleation and early growth behaviour of fatigue cracks is controlled primarily by the maximum applied stress (rather than the applied stress range), because of its marked influence on the threshold stress intensity factor of the material.

2. The preceding discussion has illustrated how fracture mechanics concepts can be utilised together with material properties and appropriate stress analysis procedures to provide guidelines which can be followed by both railway components manufacturers and users. Similar procedures may be adopted for all railway components, including wheels, couplers, axles, bogies and rail fasteners.

3. The procedure allows the design of components to be made on a technical basis rather than on historical experience which, although important, is at best qualitative in nature. The procedure can also be used by management as a tool to quantify some of the consequences related to changes in operating practices.

4. The judicious adoption and use of such concepts will result in improved cost effectiveness of railway operations.

REFERENCES
1. Mutton, P., Epp, C. and Marich, S., 1982 "Rail Assessment", A.A.R. Conference Proceedings "Second International Heavy Haul Conf.", Colorado Springs, U.S.A., Session 30.
2. Epp, C. and Okey, R., 1982 "The Influence of Wheel/Rail Interaction on System Performance", Ibid., Session 31A.
3. Longson, B.H. and Lamson, S.T., 1982 "Development of Rail Profile Grinding at Hamersley Iron", Ibid., Session 33.
4. Rooke, D.P. and Cartwright, D.J., 1976 "Compendium of Stress Intensity Factors", Her Majesty's Stationery Office, London, England.
5. Marich, S., Cottam, J.W. and Curcio, P., 1978 "Laboratory Investigation of Transverse Defects in Rails", The I.E. Aust. Conference Proceedings "Heavy Haul Railways Conf.", Perth, Australia, Session 303.
6. Cooke, R.J. and Beevers, C.J., 1973, Eng. Fract. Mech. 5, 1061.
7. Beevers, C.J., 1977, Met. Sc. 11, 362.

15 Features of elastic rail fastening device for high speed railways and improvement of it based on the results of its use

K. WATANABE, Director, Railway Technical Research Institute; Dr. Y. SATO, Chief, Track Laboratory, Railway Technical Research Institute; K. TAKAHARA, Superintendent, NIIGATA Railway Operating Division; S. UMEDA, Senior Researcher, Railway Technical Research Institute, Japanese National Railways, Tokyo, Japan

SYNOPSIS. The basic design conception about the elastic rail fastening device for high-speed railway use in the JNR had been established in the course of designing Model 102 rail fastening device to be employed for Tokaido Shinkansen under construction. Then, Sanyo Shinkansen was opened to traffic and thereafter the slab track has come to be introduced extensively and much progress has been made in design of the rail fastening device to be used on these lines. Performance results of the device are satisfactory.

INTRODUCTION

1. Of all the components of the track, the elastic rail fastening device is expected to play a multiple role under severer conditions of service than any of the other track materials. Since the first half of the 1950's when concrete ties began regularly to replace wooden ones, the JNR has employed elastic rail fastening devices to fasten the rail to ties. Particularly in the period following the inauguration of Tokaido Shinkansen a number of improvements have taken place on the devices to serve in a high speed range of train operation, utilizing abundant experience gained in the field. This paper is a report on these process.

DESIGN OF ELASTIC RAIL FASTENING DEVICE

Design conditions

2. The conditions to be considered in designing the rail fastening device include the following requirements (ref. 1);

(1) The device should have such a structure and material strength that it can ensure a stable fastened state all the time under dynamic load acting on it. Especially in case of a long welded rail, a necessary creeping resistance and horizontal rotation resistance of rail should be provided.

(2) The device should mitigate the forces acting on the substructure below the rail, thereby contributing to reduction of track settlement and materials deterioration as well as of vibration and noise.

(3) The displacement of the rail head due to tilting should be held within necessary limit for running stability of the train.

(4) In connection with assurance of track maintenance accuracy, the device should possess sufficient adjustability to the tolerance of each part of it and to its possible wear.

(5) The device should be electrically insulated to ensure the reliability of signaling and train control systems.

(6) The device should be economical.

Procedure of designing

3. In the present procedure of designing the rail fastening device in the JNR, major items to be considered are as follows;

4. Design load (ref. 2). Rail fastening devices are installed in great quantities but their replacement is relatively easy. For this reason, the design loads on them are classified from a standpoint of fatigue considering the frequency of their occurrence.

5. The design loads currently specified for Shinkansens are listed in Table 1. Load A is one of extremely rare occurrence. Load B is one of frequent occurrence. Load C is a mean load. The allowable number of cycles is 10^5 for load A and 10^7 for load B.

Table 1. Design load for Shinkansen rail fastening device

Type of load	Line	A load (kN)	B load (kN)	C load (kN)
Wheel load	Tokaido	98	86	75
	Others	111	98	85
Lateral force	Tokaido	60	30	15
	Others	68	34	17

6. Design calculations. The rail pressure acting on the rail fastening device is calculated in accordance with the theory of elastically supported beams and the tilt angle of rail is calculated accordance with the theory of torsion. From the results of calculations the loads on the fastened members and their reactions to the loads are calculated and then the dimensions of each member are determined.

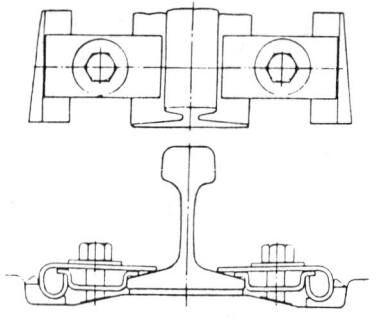

Fig. 1. Model 102 rail fastening device

Design of Model 102 rail fastening device (ref. 3)

7. Model 102 rail fastening device adopted for Tokaido Shinkansen opened to traffic about 20 years ago in 1964 had been designed also by this procedure. Fig. 1 illustrates its structure. The features of this device are as follows.

8.   Spring clip. The tip of the clip is designed to have a low spring constant of 0.5 MN/m so that variations of the rail-holding force can be minimized.

9.   Meanwhile a lateral spring is employed to spread the lateral load and mitigate its influence. The spring clip is characterized by the so-called bi-linear spring characteristics of two-point touch type (rails are fastened at two points, i.e., at the tip of the clip and at the end of rail base). Thus the fastening force can be checked in time of installation and at the same time the spring constant for vertical load can be made small, while the spring constant for rail tilting can be made sufficiently large.

10.   Spring constant of track pad. The spring constant of track pad has to satisfy two conflicting requirements that it is desirably low from the standpoints of load spreading and vibration mitigation and that it is desirably held in the range of specific values for the purpose of preventing rail tilting and lateral displacement. For Model 102 rail fastening device, the spring constant has been set low at 90 MN/m instead of 110 MN/m which is adopted for the narrow-gauge lines, after full investigations.

11.   Material quality of track pad. In the early stage of development of an elastic rail fastening device, the material of track pad was controversial on account of short service life and poor durability. The track pad for Tokaido Shinkansen was developed with an idea of replacing the track pads at the same time as that of rails. Thus an increase of the durability has been accomplished after fatigue tests such as Vibrosir test and De Mattia flexture test and the pad has been qualitatively improved.

12.   Rail fastening force. Rail fastening force is designed such that the rails are allowed thermally to elongate or shrink together with the ties. Thus the force is set at 5 kN per one spring under no vertical load (10 kN per rail fastening device), which is equivalent to a resistance to rail creep of 15 kN/m.

13.   Other details of structure. Spring supports of wedge type are employed, thereby facilitating fine adjustments of parts as manufactured and adjustments after rail wear and deterioration of members, and in consequence ensuring the accuracy of track gauge in maintenance work. As for the electric insulation, the rail contact parts which are subjected to severe conditions, other than the track pads, are not directly insulated. Instead, indirect method of insulation is adopted such as spring mount at the tail end of the fastening spring or anchor plug for bolt in concrete tie, for the purpose of avoiding deterioration of insulating materials.

CONDITIONS FOR IMPROVEMENTS IMPOSED BY HIGH SPEED OPERATION
Mitigation of impulsive force due to variation of wheel load
(ref. 4)

14.   Since 1969 the JNR has been carrying on various developmental tests with a target set at realization of commercial run on Shinkansen in the speed range of 250 km/h. In this

process a prominent wheel load variation drew attention and investigations were made to trade to the cause of such a wheel load variation under high speed run and thereby to work out countermeasures.

15.   In countermeasures, the aim had been set at making the spring constant of rail support equal to 40 MN/m per one rail fastening device. To attain this aim, it was decided on to lay ballast-mats under the ballast and at the same time to alter the spring constant of track pad from the existing 90 MN/m to 60 MN/m.

Decrease of noise and vibrations (ref. 5)

16.   Before inauguration of Tokaido Shinkansen, all the conceivable countermeasures for noise and vibrations had been taken such as introduction of long-welded rails on the entire length of the line including the bridges. Neverthless, with an unexpected increase in the number of trains operation and an increased public concern about the environmental problems as the background, noise and vibrations have come to draw attention as a social issues. To the JNR catering to the huge wayside population a solution of this problem has been urgent.

17.   Then analyses had been carried out to determine the roles and effects of track components in the vibration in the frequency domain of noise. As for the rail fastening device, it had been found important to suppress an increase in the dynamic spring constant due to the hysteresis of track pad as well as to lower the spring constant of rail fastening device.

Lateral pressure at sharp curves

18.   On a high speed railway like Shinkansen, prevention of hunting of bogie is important for securing running stability. Several preventive measures are available, but the most effective one will be to suppress the turning of the bogie. This is done on Shinkansen by supporting a part of the vehicle weight laterally at the side of the bogie frame so that a frictional resistance may work effectively.

19.   The lateral pressure to turn the vehicle on a curve will thus be increased. Such an increase in the lateral pressure will cause no trouble on a mild curve of the main line, but the increase on sharp curves will be so large that modification of the rail fastening device is deemed necessary on such curves.

Introduction of slab track (ref. 6)

20.   When Tokaido Shinkansen was extended to Sanyo Shinkansen following the completion of it in 1964 and a nation-wide Shinkansen network plan was conceived, the national status of manpower demand and supply dictated introduction of slab track. Then, for the spring constant of the rail fastening device, it was found necessary to set it at 60 MN/m, which was lower than the combined spring constant of 69 MN/m per one rail fastening device based on 90 MN/m, the value for the device on Tokaido Shinkansen and 300 MN/m for the ballast. It was found equally necessary to provide for positional adjustability of rails in vertical and lateral directions.

KEY POINTS AIMED AT IMPROVED DESIGN

Design of Model 102 rail fastening device for high speed

21.   The rail fastening device for the Sanyo Shinkansen has
been re-designed to have a spring constant of 60 MN/m based on
the consideration described under paragraph 15. Further the
load acting on the device is somewhat increased on account of
the axle weight having increased from 150 kN to 170 kN as shown
in Table 1. The designing has been successful as the result of
increasing the thickness of the fastening spring from the existing 5 mm to 6 mm and replacing the existing track pads with ones
of 60 MN/m after full deliberations and in consequence the lateral displacement of rail due to tilting is now short of 4 mm.

Improvement of track pad

22.   Durability. A decrease in the spring constant of track
pad will lead to drop in the strength of the pad on account of
an increased strain in the pad. The JNR sets the allowable
limit broadly at 2.0 MPa in mean compressive stress at center of
rail bottom and 4.0 MPa in maximum value at pad end or 10 % in
mean strain and 20 % in maximum value. In the case of the track
pad with spring constant 60 MN/m, the thickness of the pad has
been increased from the existing 7 mm - 180(length)x140(width)mm
- to 10 mm to reduce the strain, thereby preventing the fatigue
strength from dropping.

23.   Compressive deformability. The track pad, when compres-
sively loaded, is likely to exhibit non-linearity under the
influence of its visco-elasticity and under the effect of its
profile, say, groove, which has been given to it to ensure the
specified spring constant, and thus it is likely to trace a
hysteresis loop. In consequence the effective spring constant
of the track pad will have a large value in the range of high
loads and small variations of load. To reduce the non-
linearity, it will suffice to reduce the stress, if the material
is the same, and for this purpose the dimensions have only to be
enlarged. On some of slab tracks and on the later developed
other track, a large-size track pad of 270x140x10mm has been
introduced exploiting the above feature.

24.   Dynamic characteristic (ref. 7). Dynamic loading tests
have been performed on various types of track pads. The
findings from these tests are that the dynamic spring constant
becomes larger with an increased frequency, an increased center
load and with a decreased load amplitude; and that a drop in the
temperature causes not only an increase in the dynamic spring
constant but also an increase in the increment of it due to
an increased frequency. It has been revealed by these tests
that at 2500 Hz the dynamic spring constant is as high as
200 MN/m against the static one 60 MN/m, that is about 3.5
times the latter or about 2.5 times the dynamic spring constant at 5 Hz.

Designing of rail fastening device for slab track

25.   Model 8 for Tohoku and Joetsu Shinkansen. As the rail
fastening device for slab tracks, initially Model 5 which is a
direct-fastening type was designed and employed on Sanyo
Shinkansen and thereafter Model 7 of the same direct-fastening
type with a greater adjustability has been introduced (ref. 8).

Practical use of Model 5 suggested amelioration of this model
such as better stability of spring in adjustment. For this
reason, on Tohoku and Joetsu Shinkansen Model 8, i.e., an
improved version of Model 5 with modified spring clips was
introduced and has been satisfactorily serving on these lines
up to this time. Fig. 2 illustrates Model 8, whose features
are to be described below.

26.   Satisfactory adjustability. To fully ensure the precision of rail installation, the rail fastening device must be
adjustable enough. The adjustability of Model 8 is 30 mm in
vertical direction and 10 mm in lateral direction. Further,
this is exchangable with Model 7 which is adjustable up to 50mm
in vertical direction and 30 mm in lateral direction.

27.   Development of packing for
adjustment. For full exploitation of the effect of elastic
fastening on solid bed, the track
pads should serve in a state of
being pressurized all the time.
To this end, adjust-packings to
fill the gap between track pad
and underlying member have been
developed. There are two types
of them; a variable pad which
consists of a nylon bag to be
inserted beneath the track pad
and forcibly charged with syn-
thetic resin; and Hot-Melt-
Packing (HMP) (ref. 9) which

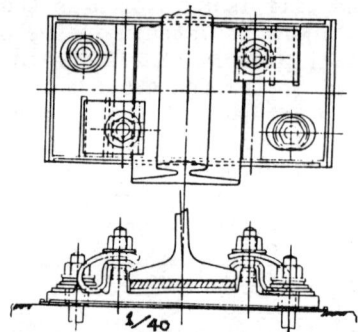

Fig. 2.   Model 8 rail
          fastening device

consists of a weave-heater-equipped resin board with a specified
thickness which is installed after lifting the rails and then
imposed with an electric current to melt and deform the resin
to a required thickness for adjustment, whereupon the rails are
lowered. The former is mainly intended for tangent section and
the latter is mainly intended for curves and for thermally
elongating and shrinking section.

28.   Regulation of resistance to rail creep. On dirct-
fastened tracks such as slab track built on structures, the
working resistance to rail creep is set at 5 kN/m per rail
which is about 1/3 of the value on ballasted track, so that an
overload may not act due to a temperature change between the
structure and the long-welded rails. This value of the resistance is realized by reducing the coefficient of friction
through use of track pads bounded on top with a stainless steel
plate as well as by reducing the rail-holding force. This type
of track pad has another merit in that it is less liable to
shearing deformation.

29.   Insulating performance. In the rail fastening device for
slab track, of which the condition for drainage is less favorable than in ballasted track, the rail fastening device must be
imparted with a high resistance to insulation breakdown and the

standard value of leak conductance for the device is set at
0.2 S/km.
30. Rail fastening device for sharp curve. On sharp curves
in urban zones a newly developed device for sharp curve is
introduced.

EXPERIENCE ON TOKAIDO AND SANYO SHINKANSEN
For ballasted track
31. Fig. 3 (a) summarizes the replacement records of rail
fastening devices for ballasted track on Tokaido and Sanyo
Shinkansen.
32. According to the results of a deterioration survey of
Tokaido Shinkansen after 10 years of its operation, the fastening springs were worn at the tip; a permanent set appeared at
midportion of the spring; and a partial erosion of the material
was recognized where the environmental conditions were un-
favorable (ref. 10). 'However, no extreme functional deterioration of the device was observed on the whole, the rate of the
repair and replacement being a little short of 1 % of the total
number of devices installed.
33. The results of investigations related to the variation
of spring constant in track pads in continuous service since
installation showed that
its value increased from
90 MN/m at first to 150
MN/m in average after 10
years (ref. 11). If a
20 to 25 % drop of
structural strength in
the track deterioration
is taken provisionally
as an allowable limit in
considering the life of
track pad, it will be
possible, according to
our track deterioration
theory, to accept a
spring constant of track
pad up to 150 to 180 MN/m.
34. Summing up the
above results, it can be
said that for the Model
102 rail fastening
device, its service life
of more than 10 years,
which is the original
target of design, has
been verified and this
device can satisfacto-
rily serve under high
speed operation.

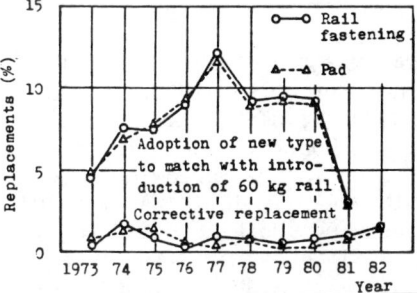

(a) Model 102 on Tokaido and
    Sanyo Shinkansen

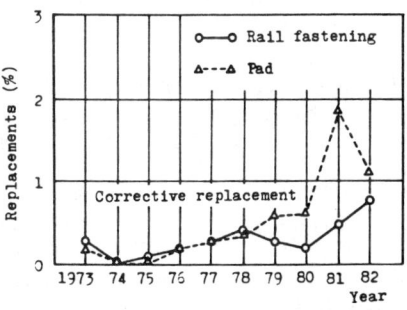

(b) Model 5 on Sanyo Shinkansen

Fig. 3. Replacement records of rail
         fastening devices and pad

For slab track
35.   After the test for practical use of the slab track on the
section Shin Osaka-Okayama of Sanyo Shinkansen, the slab track
was introduced on 69 % of the section Okayama-Hakata of it.
36.   As evident from Fig. 3 (b), the replacement rate of rail
fastening devices and track pads after inauguration of the line
has been nearly less than 1 % of the total number installed.
Thus the slab track possesses a durability far surpassing the
orginal design target.
37.   In light of experience with Sanyo Shinkansen, slab track
has been laid to an extent of 90 % on Tohoku Shinkansen and to
an extent of 94 % of on Joetsu Shinkansen.

CONCLUDING REMARKS
38.   The rail fastening devices in the JNR have witnessed
satisfactory service for over 20 years on Tokaido Shinkansen
operated at 210 km/h with 138 trains running one way per day and
38 million tons carried per year. The experience is enriched
with further service on Sanyo, Tohoku and Joetsu Shinkansen.
They have been improved as the most multi-functional component
of track for high-speed railway and are still being improved
even to this day.

References
1. WATANABE K. Engineering of rail fastening. Japanese Railway Engineering, 4, 1980, 19, 13-18.
2. SATOH Y., OTSUKI T. Durability of elastic rail fastening
devices. Quarterly Reports of Railway Technical Research
Institute (in the following shown as "Q.R."), 4, 1964, 5, 31-35.
3. MINEMURA Y., ICHIKAWA S. A101 type rail fastening devices
(the New Tokaido Line standard type) for P.S. concrete sleeper.
Q.R., 4, 1964, 5, 35-37.
4. SATO Y., SATOH Y. Cause and effects of wheel load variation
on the high speed operating line. Railroad track mechanics and
technology, Pergamon press, London, 1978, (Proceedings of a
simposium held at Princeton University, 1975).
5. SATO Y. Study on high-frequency vibrations in track operated with high speed trains. Q.R., 3, 1977, 18, 109-114.
6. WATANABE K., TAKAHARA K., SATO Y. Civil engineering maintenance of high speed railways - Track technology of high speed
railways. The paper presented in this symposium on 11 July,
1984.
7. KANAMORI T. A few experiment on elasticity of rubber pad.
Q.R., 4, 1980, 21, 187.
8. MIYAMOTO T., WATANABE K., AOKI M. Development of unconven-
tional track by JNR. Rail International, 3, 1975, 6, 189-203.
9. SHIMIZU K., YOSHIDA H. Development of packing HMP for track
levelling. Q.R., 1, 1977, 18, 15-19.
10. UMEDA S., HIRAI T. An investigation on the degradation of
spring clips used on Shinkansen. Q.R., 4, 1979, 20, 177-178.
11. UMEDA S., SAWADA T. An investigation on the deterioration
of rubber pads used on Shinkansen. Q.R. 4, 1978, 19, 177-178.

16 Rail fastenings for heavy haul railway track

T. P. BROWN, BSc(Eng), MIMechE, Technical Director, Pandrol Ltd,
London, UK

SYNOPSIS. The Paper outlines conditions of terrain, climate
and traffic in which Heavy Haul track must operate, and
current rail to sleeper fastening types are noted. Areas of
likely development are discussed, with particular reference
to providing greater attenuation of rail to sleeper dynamic
forces and increased mechanisation of fastening installation.

INTRODUCTION

1. The phrase "Heavy Haul" (HH) probably came into promi-
nence with the first Heavy Haul Conference, held in Perth,
Western Australia in 1978, and referred to purpose built
mineral hauling railways; it has since become more generally
applied to any railway operations with a predominantly unit
train freight haulage mode.

2. HH operates in some of the world's most difficult conditions of terrain and climate, with rail temperatures up to
75 degrees C in North West Australia, down to -50 degrees C in
Canada, and with annual ranges of up to 80 degrees C. Trains
can be of 250 vehicles giving a trailing weight of some 30,000
tonnes and train lengths of more than 2 kms, with track curvature of 220m and grades of 2% (Fig.1).

3. Table I details some of the railway operations which
are generally accepted as fitting into the HH category - the
Burlington Northern is included as representative of those
U.S. railroads with an important mineral haulage business.
The combinations of
heavy axle loads, high
annual tonnages and - in
some cases - severe
track geometry give
these Railways uniquely
difficult track condi-
tions.

4. Particular problems
experienced with HH track
have been detailed by
Curlewis (Ref.1) and
Oliveira (Ref.2) and can
be summarised as:-

Fig.1. Heavy Haul Track

RAILROAD	TRAFFIC	GAUGE (MM)	MAX.CUR-VATURE (METRES)	MAX.SPEED (KM/HR) LOADED	AXLE LOAD TONNES	TYPICAL RAIL (KG/M)	TRACK SLEEPERS	CONSTRUCTION FASTENINGS	ANNUAL TONNAGE (MGT)
AFRICA									
RICHARDS BAY COAL LINE (SATS)	COAL & MIXED	1065	500	60	20/26	57	CONCRETE	PANDROL/ FIST-BTR	48
SISHEN-SALDANHA BAY (SATS)	IRON ORE	1065	870	70	26	60	CONCRETE	FIST-BTR	20
LAMCO, LIBERIA	IRON ORE	1435	500	60	30	66	WOOD	ELASTIC RAIL SPIKES, DE, PANDROL	13
COMINOR	IRON ORE	1435	1000	50	25	54	STEEL	CLIP & BOLT	8
AUSTRALIA									
HAMERSLEY IRON	IRON ORE	1435	390	70	30	68	WOOD/CONCRETE	PANDROL	64
MT.NEWMAN MINING	IRON ORE	1435	580	65	30/32.5	66	WOOD/STEEL	PANDROL/TRAK-LOK	50
BRAZIL									
CVRD	IRON ORE	1000	146	60	23	68	WOOD	DE/SOME PANDROL	50
RFFSA (CENTRAL LINE)	IRON ORE & MIXED	1600	180	40(MIN.) 60(PASS.)	30	68	WOOD	PANDROL	45
N.AMERICA									
BURLINGTON NORTHERN	COAL & MIXED	1435	220	75	30	68	WOOD	CUT SPIKES (SOME PANDROL IN CURVES)	50
CANADIAN NATIONAL	COAL,GRAIN MINERALS	1435	220	80	30	68	CONC. IN CURVES ≤870M	PANDROL ON CONCRETE	45
CANADIAN PACIFIC	COAL,GRAIN MINERALS	1435	160	80	30	68	WOOD CONC.ON TEST	CUT SPIKES,SOME PANDROL PANDROL-SAFELOK	50
QNS&L	IRON ORE	1435	220	55	30	66	WOOD CONC.ON TEST	CUT SPIKES PANDROL,SAFELOK,SIDE-WINDER,LINELOC ON TEST	60
EUROPE									
RHENISH LIG-NITE MINES	COAL	1435	670	37	31.3	60	WOOD	VOSSLOH SK1/DE	-
SCANDINAVIA									
KIRUNA-NARVIK SJ/NSB	IRON ORE	1435	300	50	25	54	WOOD CONC.ON TEST	HEYBACK(PANDROL ON TEST) PANDROL	30

TABLE 1. HEAVY HAUL RAILWAYS DATA

(1)	Rapid rail wear.	(2)	Corrugation of rails.

(1) Rapid rail wear.
     Rail end batter and
     dipped joints.
     Cracked rails.
     Rapid deterioration of S&C work.

(2) Corrugation of rails.
     Sleeper degradation.
     Rapid loss of track
     geometry.
     Overturning of rails.

Of these, the second group can be considered particularly to
interrelate with the rail to sleeper fastening.

TRACK STRUCTURE

5. The track structure comprises rails, sleepers, fastenings, ballast and formation but only sleepers will be considered in this Paper in relation to fastenings.

Sleepers

6. The majority of HH track is laid with timber sleepers
(mostly hardwoods). These suffer from both mechanical
problems associated with high loadings, for example, abrasion
under baseplates and enlarging of the holes for spikes or
screw fastenings and, in some areas, severe splitting and
attacks from insects. Sleeper life can be as little as
seven years and many HH operators are evaluating the use of
alternatives of concrete and steel.

7. As far as is known all concrete sleepers in regular use
and most of these being tested are of the monoblock type,
either pre or post-tensioned. The use of concrete sleepers
is on a relatively limited (but growing) scale and results are
now generally considered to be technically satisfactory with
good installation practice and prompt vehicle and track maintenance. Their more extended use is now governed primarily
by economics.

8. Steel sleepers are not used extensively but Table I
mentions the Mt. Newman Railway which has a programme to replace wood sleepers with steel and the Cominor Rly. in
Mauretania. They are widely used in heavy duty industrial
trackwork.

Fastenings

9. Although many types of rail to sleeper fastening are
available (Ref.3), it will be seen from Table I that relatively
few are in use in HH track. Those fastening types associated
with particular forms of sleeper are listed below:-

Wood Sleepers	Concrete Sleepers	Steel Sleepers
Cut Spikes and Anchors	Pandrol	Pandrol
Pandrol	Fist-BTR	Trak-lok
DE	Lineloc	Clip and Bolt
Heyback	Safelok	
Vossloh	Sidewinder	

10. By far the most common wood sleeper fastener must be the
cut spike and rail anchor. Problems experienced are lifting
and lateral movement of the spikes giving poor gauge, and in
ensuring that all the rail anchors continue to bear against
the sides of the sleepers to resist rail creep.

11. The remaining wood sleeper fastenings are "indirect"
types in that attachment of the rail to the plate is indepen-

dent of the attachment of the plate to the sleeper. Rail to sleeper fixing is by means of screwspikes or "Lockspikes".

12. The 'Pandrol', DE and Heyback fastening systems have self-tensioning spring clips whereby a clamping force (or toe load is applied to the rail foot by the action of driving the clip into its baseplate housing.

13. Baseplates are generally rolled steel sections cut to length, and the clips of hardened and tempered spring steels.

14. Indirect resilient systems are increasingly being introduced where the most difficult operating conditions apply. Their advantages in reducing sleeper galling, gauge widening, rail overturning and rail creep, whilst making rail changing much easier, have been demonstrated; in addition, studies indicate that the additional (currently in the USA) investment of $29,000 per km can produce a discounted return of $50,000 per km, in curved track.

15. All the concrete sleeper fasteners embody a housing forming an integral part of the sleeper, and a self-tensioning spring clip located in the housing.

16. Technical characteristics of the fasteners are similar in many respects, Table 2 showing rail clamping force (toe load) and rail creep resistance values; choice is very much guided by service experience and practical features such as ease of application and re-usability after, for example, changing rails.

FASTENING	TOE LOAD (RAIL SEAT)	CREEP RESIS. (RAIL SEAT)
'Pandrol'	* 24.5 KN	* 12.6 KN
Fist	* 25.0 "	* 12.4 "
Lineloc	+ 26.7 "	+>10.7 "
Safelok	+ 18.0 "	+ 14.0 "
Sidewinder	+ 26.7 "	+ 15.6 "
*SATS Tests +Manufacturers Data		
Table 2. Fastening Properties		

17. As use of concrete sleepers has extended into difficult operating areas, clip toe loads have been increased to combat rail creep and sleeper skewing. Nominal clip toe loads per rail seat are now in the order of 20-25 KN (Table 2).

18. The performance of many rubber rail pads in early HH installations - where the pads suffered from abrasion, cutting and permanent set - has led to the use of materials with greater durability such as Ethyl Vinyl Acetate (EVA), High Density Polyethylene and Polyurethane.

19. This has been accompanied by an increase in the load/deflection stiffness of the rail pads, reducing rail movement and minimising abrasive wear on clips and insulators.

20. Insulating pieces used in the 'Pandrol', Lineloc, Safelok and Sidewinder assemblies are intended to be easily replaceable items which will wear in preference to the cast-in shoulder component (the Fist fastener incorporates insulation in a different manner). In many cases, however, insulator life has been unacceptably short, particularly in curves of radius below 6 degrees (290 metres). Intensive development has now produced designs which are giving much better performance in some of the most severe conditions known for concrete sleeper use in HH track, with curves down to 11 degrees (160

metres) carrying annual tonnages of 55 mgt on a single track.

21. The fastening systems used in HH with steel sleepers
(Table I) are 'Pandrol' and Traklok (both of the self-
tensioning type), and rigid clip and bolt designs. The first
two have rail/sleeper insulation, the development of satisfactory insulation being a major factor in any more extensive use
of steel sleepers (Ref.4).

LOAD ENVIRONMENT

22. Rail fasteners require to resist the following major
forces:-

Vertical

23. Whilst much work has been carried out (Ref.5), (Ref.6),
(Ref.7) to determine wheel to rail forces in HH conditions
less data is available as to the forces acting on individual
sleeper rail seats.

24. Fig.2 gives one example of vertical load measurements
from N.American routes from which it will be seen that there

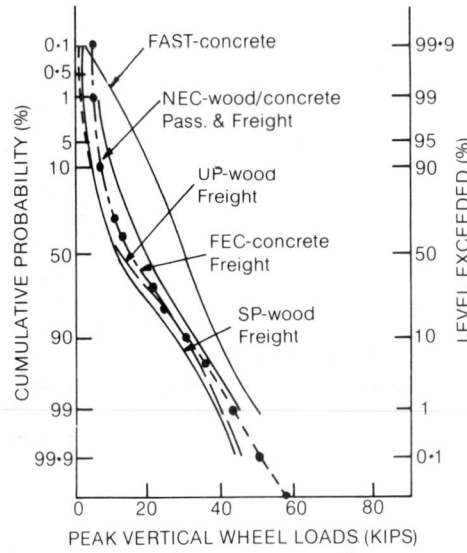

is a 0.1% probability
of a peak wheel/rail
load of 222 KN (nominal
wheel loads ≤147 KN).
With the generally
accepted form of dist-
ribution to individual
sleepers this indicates
peak loads of 116 KN per
fastening for sleepers
at 620 mm spacing
(typical of concrete
sleepers) and 89 KN
for sleepers at 495 mm
spacing (typical of
wood sleepers) (Ref.8).

25. Nominal HH axle
loads do not generally
exceed 30 tonnes,
although there are plans
by some operators - Mt.
Newman Mining for
example - to increase

Fig.2. Wheel-Rail Load Spectra.

nominal axle loads to 32.5 tonnes and perhaps 35 tonnes eventually (Ref.9). There is also limited use of such higher
axle loads by some U.S. railroads.

26. More serious effects on fastenings, however, arise from
impact and vibration generated by wheel and rail head imperfections, particularly if combined with high train speeds.
Impacts generated by wheel flats or "out-of-round" wheels, for
example, can increase static wheel load values by a factor of
three (Ref.10). The effects of rail head defects such as
wheel burns, corrugation and imperfect welds can be equally
severe - albeit more localised - and considerable effort is
now put into rectification of such defects by, for example,

grinding the running surface of the rail.

27. Fastenings are also subjected to vertical uplift forces which are not considered likely to exceed 5% of the vertically downwards fastener load (Ref.11).

28. Vertical loads on concrete sleeper fastenings affect mainly rail pad life by abrasion (and possibly internal heat build up in elastomeric types), and clip fatigue life if dynamically soft rail pads are used. On wood sleepers fatigue failure of baseplates can occur if not adequately supported by the sleeper surface.

Lateral

29. Lateral forces appear to become significant in curved track when the radius is below 5-6 degrees (350-290 m). There is less data available as to the levels of lateral force arising in HH curved track but wheel to rail (high rail of curve) forces of 89-111 KN have been reported with extreme values reaching 178 KN, although of very short duration (Ref.12). Tests on CP Rail have given measured values (in 220 m curve) of 61 KN (Ref.13) and Prause (Ref.11) suggests that frequent lateral forces of 40% of the vertical load with 60% as maximum are reasonable estimates. Analysis (Ref.14) has indicated that 80-90% of the lateral wheel/rail force can be carried by a single fastener.

30. On concrete sleepers the tendency of the rail to rotate ("roll-over") leads to uneven pressure distribution on the rail pad, with consequently increased tendency to distortion and abrasion and, in extreme cases, cutting of the pad by the edge of the rail foot. Insulators are subject to very high crushing forces between rail foot and the gauge retaining component cast into the sleeper. Resilient rail clips at the gauge side of the rail are given additional dynamic deflection which in extreme cases can result in loss of clip toe load or fatigue failure.

31. On wood sleepers lateral forces accentuate the abrasion of the sleeper surface under the baseplate (plate "cutting") and can distort and lift cut spikes. With resilient clip fasteners high lateral forces could cause greater rail rotational movement and thus clip dynamic deflection and fatigue failure. This does not seem to be a problem in practice; probably because of the uniform vertical restraint on the rail foot achieved with modern resilient fastenings.

Longitudinal Rail Restraint

32. High tractive and braking forces, and large temperature ranges, found in HH operation all lead to a tendency for rail to move longitudinally. Resilient fastenings give a uniform resistance to rail movement at each rail seat of the order of 12 KN (in static tests), this can be reduced by 30-40% if vibration is applied to the rail (Ref.15) (Ref.16). To minimise rail pull-apart in low, and track buckling in high, temperatures, the AREA (Ref.8) requires a minimum static creep resistance of 10.7 KN per fastener. Very high values of creep resistance are unnecessary and undesirable since they can cause sleepers to plough through the ballast.

FUTURE DEVELOPMENTS

33. The trend to higher clip toe loads, which has been the most immediately effective way of improving fastening performance in HH conditions, has been accompanied, on concrete sleepers, by the use of stiffer rail pad materials. Recent research (Ref.17) (Ref.18) has indicated, that high values of stiffness in the connection between rail and sleeper can have detrimental effects on concrete sleepers as a result of the transmission of impact and vibration via the fastening system. Although centred primarily on the operation of high speed trains with imperfect wheels, it is thought that this current work directed towards improving the impact attenuation properties of concrete sleeper fastenings will be reflected in future designs of HH fasteners. Most current effort primarily involves the rail pad but optimum performance will only be obtained by consdiering the total track structure in determining the optimum dynamic performance of individual components. Any such change is dependent upon identifying materials for rail pads and insulating components of HH fasteners which will also have resistance to abrasion and permanent deformation not readily found in presently known materials.

34. The provision of a connection which will give greater dynamic isolation between rail and concrete sleeper will assume even greater importance if cost pressures lead to any reduction in the generally good standards of wheel and rail maintenance practised by most HH railways, or any increase in train speeds.

35. Mechanised installation of fasteners is assuming greater importance with the need to cut costs, unwillingness to work unsocial hours and the need to apply fastenings in step with mechanised methods of tracklaying. This trend may well lead to new fastenings designed specifically for machine application, but for the more immediate future effort is concentrated on mechanising the application of existing fastening types.

36. Two types of machines for driving fastenings are under general development. The most common covers machines for driving manually placed clips. The latter type of machine is designed for fully automatic sleeper finding, insulator and clip placement and clip driving.

CONCLUSION

37. Whilst the forces to be contained by HH fastenings may not increase appreciably over the next decade greater knowledge and understanding of these forces and resulting track dynamics will lead to changes in fastening characteristics, with attempts to introduce more attenuation of rail to sleeper forces whilst retaining reliability and subject to cost constraints. This is of particular importance if the economic case for particular operations substantiates the wider introduction of concrete sleepers.

38. Cost pressures will also lead to growing mechanisation of fastening installation and the development of fastening systems to suit such installation.

1.   CURLEWIS Col.W.P.C.   18,000 tonne ore train in Australia. Institution of Mechanical Engineers, paper, 1974.

2.   OLIVEIRA Dr.J.H.S.   Special rails on Heavy Haul Railways. Heavy Haul Railways Conference, Perth, Western Australia, September 1978.

3.   BROWN T.P.   Development of rail fastening systems to meet changing requirements.   Proceedings, AIT Symposium, Madrid, November 1981.

4.   BROWN Dr.J.H.   Design refinements make the steel sleeper viable.   Railway Gazette International, October 1979.

5.   C.N.Rail Research Centre, Montreal, october 1979. Traffic load spectra, peak vertical wheel - rail loads.

6.   ZAREMBSKI A.M. and ABBOTT R.A.   Fatigue analysis of rail subject to traffic and temperature loading.   Heavy Haul Railways Conference, Perth, Western Australia, September 1978.

7.   PRAUSE R.H., et al.   An analytical and experimental evaluation of concrete cross tie and fastener loads.   Federal Railroad Administration Report FRA/ORD - 77/71.

8.   AREA Bulletin 634.   Preliminary specifications for concrete ties (and fastenings).

9.   MAIR R.I. and MARICH S.   Heavy Haul tracks may accept 35 tonne axle loads.   Railway Gazette International, August 1983.

10.   HARRISON H. and MOODY H. Correlation analysis of concrete cross tie track performance.   Proceedings, Second International Heavy Haul Railway Conference, Colorado Springs, U.S.A., September 1982.

11.   PRAUSE R.H. et al.   An analysis of performance requirements for cross ties and fasteners.   Federal Railroad Administration Report FRA/ORD - 78/37.

12.   MARTA HA.A. and KOCI L.F.   Wheel and rail loadings from diesel locomotives.   Proceedings of the Railway Fuel Operating Officers Association 1971.

13.   GONSALVES R. et al.   Truck evaluation test 'TP 2970'. Department of Research, Canadian Pacific Limited Report No. 5670-81, March 1981.

14.   MEACHAM H. and PRAUSE R.   Studies for rail vehicle track structures.   Federal Railroad Administration Report No. 1 FRA-RT-71-45, April 1970.

15.   WILDENBOER L.A.   Report on performance of 'Pandrol' sleeper fastenings on the railway network of South African Transport Services.   Track Development Report No. 24644/01/83.

16.   LUNDKVIST B. and ASTROM L.   Swedish State Railways Report No.8, April 1975.

17.   MOODY H.G.   Some aspects of concrete tie performance in FAST and in revenue service.   Proceedings - FAST Engineering Conference, November 1981.

18.   DEAN F.E. et al.   Effect of tie pad stiffness on the impact loading of concrete ties.   Proceedings Second International Heavy haul Railway Conference, Colorado Springs, U.S.A., September 1982.

(Note: 'Pandrol', DE, Fist-BTR, Heyback, Lineloc, Safelok, Sidewinder, Trak-lok are Trademarks.)

206

Discussion on Papers 13–16

MR D. F. CANNON, Senior Principal Scientific Officer, British
Rail Research and Development Division, Derby

It is appropriate that there has been a paper concerning
fatigue.

Mr Purbrick has said that the costs incurred as a
consequence of unexpected rail failure are considerable. He
also made it clear that on British Rail's high grade, high
revenue track it is fatigue, not wear, that is dictating the
rail removal policy.

Mr Campbell has said that in the future we should be
concerned with increased reliability and that the customer
demands reliability as much as speed.

I cannot recall one occasion when corrugations have
resulted in derailment and timetable disruptions –
unfortunately I can recall too many occasions when rail
fatigue has resulted in these problems.

Mr Cooper's presentation indicated that railways find it
difficult to quantify the cost of damage caused by
corrugations. However, we do know the costs of rail
fractures.

I therefore suggest that it is fatigue which is likely to
dominate the reliability of rail in a mixed railway
environment.

With regard to Mr Marich's paper, which is in principle a
reasonable approach to one aspect of rail fatigue, there are a
few problems.

(a) At short crack lengths (1 mm or less) simple fracture
    mechanics ceases to be valid and it is necessary to
    invoke short crack fracture mechanics theories which
    today are only in their infancy.
(b) It is asking a great deal of ultrasonic testers to detect
    and size cracks of 1 mm and less depth.
(c) In very high wear conditions, e.g. in tight curves,
    fatigue may not be too much of a problem since rail life
    is short anyway.

Finally, fracture mechanics has helped to understand rail

fatigue problems. For example the significance of stresses is
now more fully appreciated and this has provoked studies of
changing steel mill roller straightener procedures and
investigations into new methods of rail straightening (e.g.
stretching) which could lead to new rails containing less
damaging residual stresses.

DR A. ZAREMBSKI, Director of Research & Development, Pandrol
Incorporated (USA)

Early work on rail fatigue life put significant emphasis on
three major types of stress:

Hertzian contact stresses $\Big\}$ from dynamic load environment

Vertical rail bending

Longitudinal thermal stresses in an ambient temperature
environment.

Subsequent work has downgraded the effect of vertical bending
and longitudinal thermal stresses but has introduced a
significant effect due to rail residual stresses.

Mr Marich, given the apparent success Australian mining
railroads have had with controlling rail fatigue with
metallurgy and profile grinding (shifting the loading point),
what do you feel are the dominant loading (and stress)
contributions to rail fatigue failure, particularly under
heavy axle load conditions?

DR -ING J. EISENMANN, Professor of Civil Engineering,
Technische Universität München

ADDITIVE STRESSES IN RAILS
The concentrated load applied to the rail head as well as the
constriction of the rail cross-section lead to an interference
with stress propagation in the area of the rail head [1, 2].
When the load is applied centrically to the rail head, the
rail web undergoes a compression which deviates from bending
theory. This interference with the propagation of the bending
stress is shown in Fig. 1. An approximate calculation of this
interference stress can be performed by means of the
computational method given in reference 2. The bending
pressure stress is increased at the upper side of the rail
head and is decreased at the lower side of the rail head. At
the lower side of the rail head, smaller bending tensile
stresses appear. The interference stresses calculated for
various rail cross-sections are assembled in Fig. 1. They
agree well with the experiments that have been performed.

When the wheel load is applied off centre at the rail
head, and when a lateral force acts simultaneously, the rail
is first stressed by torsion and, in the horizontal direction,

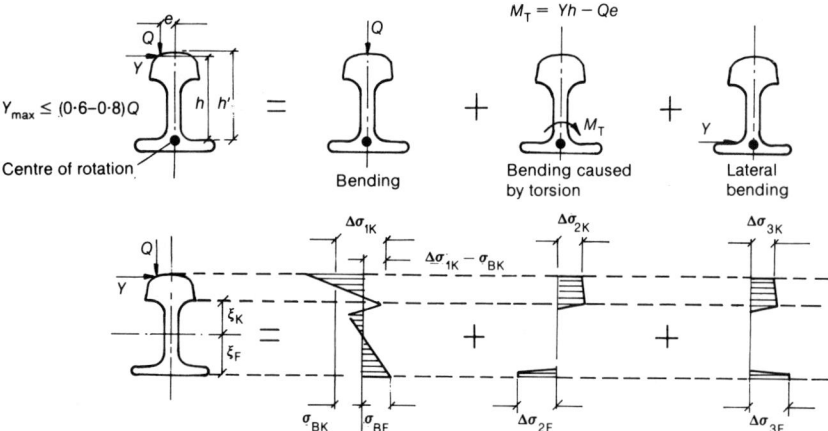

Fig. 1. Decomposition of forces and propagation of stress in the rail for an eccentric vertical load and a horizontal load

Coefficients for calculating the bending stress at the lower edge of the rail head and at the outer edge of the rail base

$$\sigma_{BK} = \frac{M_B}{I} \times \xi_K = \lambda_{OK} \times M_B \times 10^{-8} \qquad \sigma_{BF} = \frac{M_B}{I} \times \xi_F = \lambda_{OF} \times M_B \times 10^{-8} \quad (N/mm^2)$$

$$\Delta\sigma_{1K} = \lambda_{1K} \times Q \times 10^{-5} \qquad \Delta\sigma_{1F} = 0 \qquad (N/mm^2)$$

$$\Delta\sigma_{2K} = \lambda_{2K} \times M_T \times 10^{-8} \qquad \Delta\sigma_{2F} = \lambda_{2F} \times M_T \times 10^{-8} \qquad (N/mm^2)$$

$$\Delta\sigma_{3K} = \lambda_{3K} \times Y \times 10^{-5} \qquad \Delta\sigma_{3F} = \lambda_{3F} \times Y \times 10^{-5} \qquad (N/mm^2)$$

where Q and Y are in newtons and M_T is in newton millimetres

Calculation of the stresses at the edge of the rail head and at the edge of the rail base

Rail profile	S 49	S 54	S 64	UIC 54	UIC 60
I (mm^4)	1819×10^4	2073×10^4	3252×10^4	2346×10^4	3055×10^4
W_u (mm^3)	248×10^3	276×10^3	403×10^3	313×10^3	377×10^3
h (mm)	99	102	124	119	129
At the lower edge of the rail head					
λ_{OK}	200	170	150	200	180
λ_{1K}	54	49	48	58	50
λ_{2K}	1600	1400	1300	1900	1700
λ_{3K} *	160	140	100	120	110
At the outer edge of the rail base					
λ_{OF}	400	360	350	320	270
λ_{2F}	1520	1310	950	1240	1260
λ_{3F} *	298	261	203	240	229

* Sleeper spacing 650 mm

209

by bending. Because of the constricted cross-section, a
warping torsion appears at the point where the load is
applied. Like the horizontal bending stress, this torsion
leads to bending stresses at the edges of the rail head and
the rail base. An approximate calculation can be performed
with the computational method in reference 2. This method is
based on the work of Timoshenko and Langer [3]. For a
prescribed vertical load with eccentricity e, and for a
prescribed lateral force Y, the additional bending stress on
the edge of the rail head and the rail base can be calculated
by means of this method. The horizontal bending stress of the
rail can to a first approximation be taken as that of a
single-span beam with a support width corresponding to the tie
spacing. The force Y, which appears at the fulcrum, must be
divided in accordance with the horizontal moment of inertia of
the rail head and the rail base. The result of this
calculation for the usual rail cross-sections is also shown in
Fig. 1. The resulting values exceed those from laboratory
experiments by 10%.

In addition to the stresses arising from the traffic load
and the temperature, bending tensile stresses are generated at
the outside edge of the rail base, in tight curves, when the
rail is bent. These tensile stresses must be included in the
formulation.

REFERENCES
1. Schlumpf, U. Über eine bisher nicht berücksichtigte
   Beanspruchungsart von Eisenbahnschienen. (A previously
   unconsidered type of stress on railroad tracks.) Schweizer
   Bauzeitung (1954), No. 1, pp. 6–9.
2. Eisenmann, J. Theoretische Betrachtung über die
   Beanspruchung der Schiene am Lastangriffspunkt.
   (Theoretical considerations on the stress of rails at the
   load contact point.) Eisenbahntechnische Rundschau
   (1965), No. 1/2, pp. 25–34.
3. Timoshenko, S. and Langer, B.F. Stresses in railroad
   track. Trans. Am. Soc. Mech. Engrs, 54 (1932), applied
   mechanics section.

DR M. H. MAGUED, Morrison Hershfield Ltd, Toronto

The approach presented is very interesting and has significant
merit. The points raised by Professor Eisenmann are
rigorously correct from an engineering mechanics viewpoint.
However, significant complications would be faced in the
calculation of stresses, particularly in worn rail sections.

An empirical approach to the determination of the safe
minimum rail head section is most appropriate. Although it is
true that current fracture mechanics theories are not
applicable to carbon steels of 600–800 MPa yield strength,
previous work indicates that the concept can be used to

predict crack growth rates under fatigue loading. However, proper empirical calibration is required.

Mr Marich, could you clarify the criteria used for the establishment of the 0.5 mm crack depth threshold? Was this based on the crack size considered to be detectable? Could you clarify further the details of the test set-up and the method of load application?

MR R. GOSTLING, Head of Track Research Unit, British Rail, Derby

Dr Sato, in your paper, pad stiffnesses up to 2.5 kHz are quoted. How were these measurements carried out?

DR Y. SATO

The measurement of pad stiffnesses was performed up to 4.0 kHz including 2.5 kHz at which value I discussed the spring constant in the paper. The measurement was performed according to the well-known law of rheology for calculating the frequency characteristics from the temperature characteristics. Precise measurements for the pad were performed at temperatures of 20, 10, 5, 0, –5 and –10 $^{\circ}$C for frequencies of 5, 10, 20, 30 and 40 Hz with the use of an electro-hydraulic type of fatigue testing machine (Servo-pulser).

MR A. VAN EYKEN, Canadian Institute of Guided Ground Transport, Kingston

Mr Brown, in relation to the question of longitudinal rail restraint, paragraph 32, why is it necessary to have rail fastenings at intermediate positions to provide longitudinal restraint to cater for large temperature ranges? If these are considered in isolation without reference to traction and braking forces, then the temperature-induced rail load is uniform along its length. This means that the anchors at the two ends of the rail ribbon in question must be capable of reacting the load. This would make the stated need for intermediate longitudinal fastener restraint unnecessary for both compressive and tensile loads.

It may be argued that longitudinal restraint is necessary to prevent large separation of rails in the event of rail fracture due to low temperature. The need is not so obvious for compression loads due to high temperature effects.

Why was it thought necessary for rail fasteners to provide longitudinal reaction to forces due to temperature effects? A more important requirement is lateral support capability.

MR T. P. BROWN

Longitudinal restraint in rail fastenings is necessary, as you
indicate, in fasteners installed in the body of a length of
long welded rail if movement due to traction and braking
forces is to be prevented and if rail 'pull-apart' is to be
minimized in a low temperature rail break.

The designs of current fastening types - which have
developed as a result of trying to combine technical
requirements with the needs for simplicity and reliability in
the field - are essentially very similar in that the rail is
connected to the sleeper by a metallic spring applying a
vertically downward clamping force to the top surface of the
rail foot. The arrangement not only then generates
longitudinal restraint in the fastening assembly but also
generates resistance to rail rollover, resistance to torsional
movement of the rail about a vertical axis and ensures that
the rail remains in vertical contact with the baseplate or
rail pad/sleeper under dynamic loading - thus minimizing the
wear of components arising from relative movement.

I see little advantage in designing a fastener which
retains all these characteristics with the exception of
longitudinal restraint. Apart from the consequent need to
ensure that the correct fastener is then used (and continues
to be used) in the appropriate section of track - which goes
against the benefits of standardization and simplicity - I
believe problems would arise from traction and braking forces,
for example, affecting the stress distribution in a length of
long welded rail, with the likelihood of a lowering of the
stress-free temperature towards the 'running-on' anchored end.

I agree with your point regarding the need for lateral
support capability; this is a feature which has required much
development in applying resilient fastenings to those heavy
haul railways with severe curvature conditions but as
indicated in the paper it is now possible to provide lateral
support which is adequate for the most severe conditions in
which resilient fastenings have so far been applied in revenue
service.

MR B. I. SINGAL, Design Manager, Mass Transit Railway
Corporation, Hong Kong

In your paper, Mr Brown, you state that for heavy haul
operations clip toe loads have been increased to combat rail
creep and sleeper skewing. Later it is suggested that a
minimum static creep resistance of 10.7 kN per fastener is
required to minimize pull-apart in rail fracture or track
buckling.

In underground railways on non-ballasted track the
problem of sleeper skewing is irrelevant. Temperature changes
are low: hence the dangers of pull-apart or track buckling are
low.

There is thus a case for a significant reduction in the
clip toe load on underground railways. This will be an
advantage by improving track resilience and the life of the
resilient pad on NB tracks which use a resilient pad next to
the rail.
Do you wish to comment on this?

MR T. P. BROWN

I agree with the desirability of not applying high clip toe
loads where thermal and other forces are low. This gives
advantages both in utilizing more effectively the
characteristics of resilient rail pads and in lowering the
mean stress levels in fastenings – although not necessarily
the dynamic stress ranges. On mass transit railways, however,
although thermal forces may be low in tunnel sections, the
longitudinal forces created by braking and traction cannot be
neglected, nor can the high lateral forces which can arise in
the sharp curves of some metro systems. Clip toe load plays a
part in resisting these and is frequently determined by the
need to meet quite rigorous criteria (especially with regard
to resistance to rail rollover) in many metro specifications
for rail fastenings.

DIPL -ING HENNING VON HEIMBURG, Vossloh-Werke GmbH

The report informed us that tension springs for heavy haul
railway track can be overstressed for a time and in that case
they can break down. In our experience tension springs which
are correctly dimensioned have long outlasted the durability
of the rails. A correct construction design and correct
calculations are essential.
If you want to increase the life of pads and insulating
pieces, it is preferable to use screws in wood sleepers
instead of spikes. A permanent and positive connection can be
ensured for example with spring washers. Such a design also
facilitates mechanized installation.
Rail fastenings for concrete sleepers have no self-
tensioning spring clips. For example, the best known Vossloh
fastening systems for wood, concrete or steel sleepers are
tightened by screws and therefore are especially suitable for
mechanized installation. For such systems it is also
profitable to have a rotatable fastening for the rails, to
prevent the rails from moving in sharp curves and to ensure
that the lifting wave of the rail under the wheels does not
lead – under heavy haul operation – to an early breakdown of
the pads between the rails and the sleepers. Insulating
pieces, which are placed loosely between the rail foot and the
fastening device are always subject to high abrasion and
therefore to early crushing. For economy such fastenings
which do not have loose parts between the rail foot and the

fastening device are preferred. The Vossloh systems already
mentioned do not have loose parts in this range.

It was quite correct to point out that in future much
more attention will be paid to the costs of reconstruction and
maintenance of rails than before. A mechanized installation
is facilitated when the rail fastenings, including all
components such as pads, are completely fitted to the sleeper
at the sleeper factory: not only is there no distribution of
the components on the track but through the correct fastening
of the necessary parts on to the sleepers mechanization of the
installation is facilitated. Experiences, e.g. with Vossloh
fastening systems, have shown this for installation on
concrete sleepers as well as on wood sleepers.

The economical use of concrete sleepers will be much more
profitable if a fastening system is selected that ensures (a)
an easy adjustment of all rail sections, (b) that rail
sections with different breadths of rail foot (by fixing the
gauge) can be installed on the same sleeper and (c) that all
fastening components, including the dowel in the sleeper, are
changeable. With such a system it is possible, if necessary,
to obtain gauge variations of 10 mm on each side by replacing
only the guide plates.

MR T. P. BROWN

As mentioned in the paper most heavy haul railways have chosen
self-tensioning indirect types of fastenings when changing to
use of a resilient connection between the rail and the
sleeper. Many of the features of screw-type fastenings
mentioned by Dr Henning Von Heimburg are agreed, but such
designs require careful control when fitting, to ensure
consistent toe load, and it is generally accepted that there
is a continuing maintenance requirement if screw threads are
to remain in a condition to permit quick rail release or
adjustment of the toe load.

The point regarding the desirability of being able to use
alternative rail sections is also agreed, but again this
facility is also possible with the more widely used self-
tensioning designs.

It is a matter for judgement whether separate insulating
pieces must suffer from 'early crushing' but where lateral
forces are very high, and they must be transmitted to the
sleeper, I believe that it is desirable to have a sacrificial
component in the fastening system which will serve to protect
the more expensive items – including the concrete sleeper
itself.

17 The generation and perception of vibration from rail traffic

Dr. M. J. GRIFFIN, Institution of Sound and Vibration Research, University of Southampton; C. G. STANWORTH, Research & Development Division, British Railways Board, UK

SYNOPSIS. Wayside vibrations caused by trains on main line railways has been investigated both in field and laboratory studies. Initial results have suggested a measure of vibration dose which seems to correlate well with subjective responses to railway vibrations imposed in a laboratory. The frequency weighting and the form of integration in the "dose", when combined with features of the generation process revealed experimentally, suggest the extent and form of remedial measures.

INTRODUCTION

1. As railway vehicles pass along the track, vibration is generated in the ground. Many factors intervene in the generation process - vehicle features such as weight, speed, suspension design, dimensions, - track features such as the longitudinal profile in detail; rail joints, sleeper pitch, long wavelength corrugation. Below the track, the nature of the ground is also important, and is further significant in the propagation of vibration to the wayside.

2. When vibration from the railway passes into a building, it can suffer substantial modifications of level and spectrum (including amplification), and may be perceived by people in the building - a potential source of complaint.

3. Increasing awareness of the influence of the environment on the quality of life has resulted in the formulation of standards and limits which one might be tempted to apply to railway-induced building vibration. The characteristics of the vibration produced in buildings by surface railways differs greatly from those produced by other common sources such as domestic and industrial machinery, blasting, road traffic and aircraft. Interpretation of the standards is therefore necessary so that the appropriate weight is given to the magnitude, duration and frequency-of-occurrence of each vibration event. Without knowledge of the reactions of wayside residents to different types of motion it will not be possible to predict their response and so define the vehicle, track or operating characteristics which will minimise disturbance to the occupants of buildings.

VIBRATION STANDARDS

4. Within Britain the two most significant current standard documents are BS 6472 (ref. 1) and ISO 2631, draft Addendum 1 (ref. 2). Both indicate the magnitudes of continuous and intermittent vibration which would be 'satisfactory'. For continuous vibration in residential properties during the day a maximum value 2 to 4 times a base curve (the base curve is very approximately the threshold of perception) is considered satisfactory. At night, a value of 1.4 times the base is recommended. It is stated that doubling these suggested magnitudes for continuous vibration may result in adverse comment which will increase significantly if they are quadrupled. For day-time intermittent and impulsive vibration, values of 60 to 90 times the base curve are suggested. At night this is reduced to 20 times the base curve. Figure 1 shows the magnitude of vertical (z-axis) vibration corresponding to the base curve and to 2, 4, 8, 16, 32 and 60 times the base curve. For vibrations in the railway frequency band the horizonal (x- and y- axis) curves are similar, but at a rather higher level.

5. Since railway vibrations lie mostly in the spectral range above 8 Hz, the vibration acceleration weighting characteristic of Fig. 1 is equivalent to a constant vibration

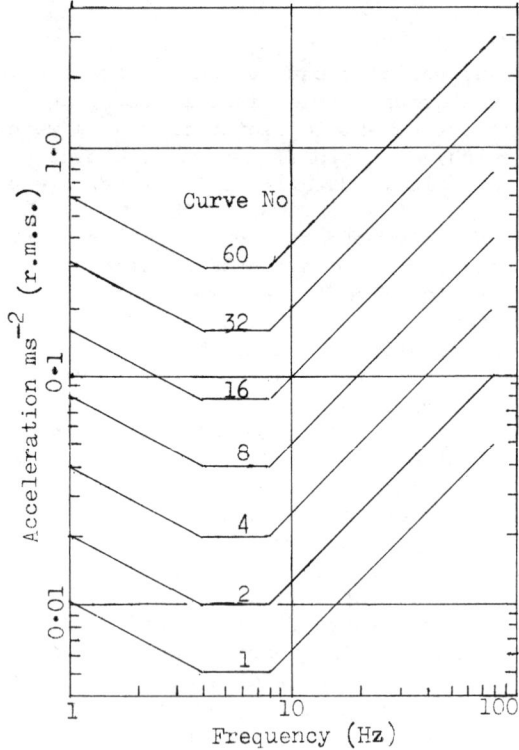

Fig. 1. Building vibration z-axis curves for acceleration (rms)

velocity, being 0.1 mms^{-1} rms for the base curve.

6. The large differences between the suggested satisfactory magnitudes for continuous vibration and those for intermittent and impulsive vibration emphasise the need to decide how the variable durations and wide range of frequencies-of-occurrence of railway-induced building vibration should be assessed. Magnitudes 60 times the base curve will not normally be acceptable while there will be instances where it is impossible to achieve values as low as 1.4 times the base curve.

HUMAN RESPONSE TO RAILWAY VIBRATION

7. Laboratory experiments have investigated the reactions of up to 40 subjects to reproductions of the vertical vibration from railways recorded in houses in the ·UK and Germany (see ref. 3). The motions varied in duration from about 9 to 27 seconds and contained frequency components in the range 8 to 60 Hz. A typical acceleration time-history and power spectrum are shown in Fig. 2.

8. The results confirmed the expectation that subject reaction could not be well predicted from solely the overall root mean square accelerations, the peak accelerations or the durations of the motions.

9. From 108 alternative measures of the vibration it was possible to select the frequency weighting and averaging procedure which gave the best correlation with subject judgement. Of those tested, the ISO z-axis frequency weighting (Fig. 1) gave the best predictions. It was concluded that no averaging time was necessary if the severity of each stimulus was determined from the vibration dose:

$$\text{Vibration dose} = \left(\int_0^T a^4(t) \ dt \right)^{1/4}$$

where a(t) is the frequency weighted vibration acceleration time history.

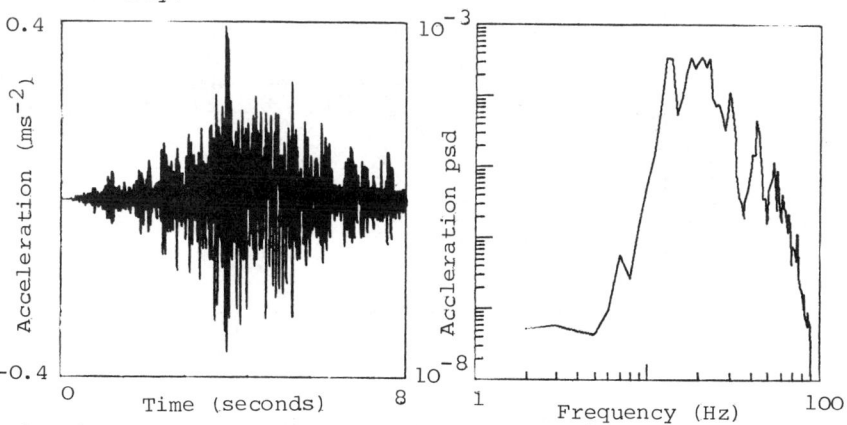

Fig. 2. Acceleration time history and power spectrum of vertical building vibration during the passing of a nearby train.

217

10. This dose procedure incorporates a weighting for the duration of each motion and the effect of the number of trains in a day. It suggests that, in a convenient example with trains producing the same pattern (duration etc) of vibration, a single train may produce vibration of double the magnitude of 16 trains or four times the magnitude of 256 trains and yet cause similar disturbance. The procedure also suggests how the influence of trains of different characteristics may be summed. Further research on the relation between disturbance, frequency of occurrence and vibration magnitude is taking place in current laboratory studies.

11. Figure 3 shows how satisfactory magnitudes of vibration are adjusted when the dose procedure is used in the context of BS 6472 and ISO 2631 DAD1. The curves indicate the satisfactory magnitudes corresponding to durations of constant magnitude vibration per day from 0.0625s to 16 hours. The values are drawn assuming a multiplying factor over the base curve of 2 for continuous vibration. They would be increased by 2 (ie raised to the next curve above) if a starting value of 4 is used. Satisfactory magnitudes for each duration may therefore be considered to be in the region above the curve for the appropriate duration and below the next curve above.

12. For perceptible railway-induced building vibration the exposure durations may be expected to be in the range 16s to

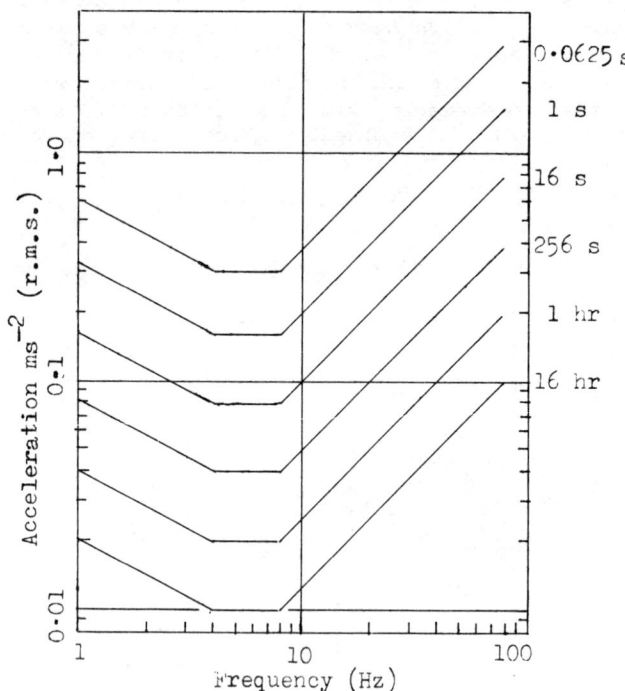

Fig. 3. Satisfactory vibration magnitudes for various dose durations per day.

about 1 hour per day. Assuming the lower multiplying factor over the base curve of 2, the range of satisfactory magnitude of frequency weighted railway-induced building vibration is then from 0.02 ms^{-2} rms (equivalent to 0.4 mms^{-1} for railway vibrations) to 0.08 ms^{-2} rms (1.6 mms^{-1}) depending on how often it occurs. Adverse comments would be expected with weighted values in the range 0.04 ms^{-2} rms (0.8 mms^{-1}) to 0.16 ms^{-2} rms (3.2 mms^{-1}) and these may be expected to increase significantly if the vibration exceeds the band from 0.08 ms^{-2} rms (1.6 mms^{-1}) to 0.32 ms^{-2} rms (6.4 mms^{-1}). A survey of the reactions of people near railways is investigating how these predicted values relate to public awareness of the vibration near railways.

13. From the way in which this dose procedure operates, it is clear that any vibration countermeasures must take into account the effects of the worst trains, and that a factor-of-improvement in the range 2 to 4 should achieve substantial benefit.

THE CAUSES OF RAILWAY VIBRATION

14. That railway vibration can sometimes be a nuisance is evidenced by the complaints received by the Board, although these are a small number by comparison with the number of people who dwell at the lineside of the some 18 000 route-km of the system. Many more people perceive vibration from trains, however, than are moved to complain. This is illustrated by the Fields and Walker survey (ref. 4) which included a few questions on perceived vibration in its questionnaire.

15. Analysis of the vibration complaints has shown that no class of vehicle seems to be completely innocent, but some vehicles are more likely to cause complaint than others. Electric multiple units (EMU) were more likely to cause complaint than diesel multiple units, and probably the higher unsprung mass due to the traction motors was responsible. Secondly, wagons with wedge-loaded friction damped suspensions (either 2 or 4 axle) carrying bulk dry cargo at 25t per axle were particularly likely to cause complaint (and were a far greater problem than EMUs). It was also clear that one or two particular trains per day could be the cause of complaint, even when they were a tiny proportion of the whole service.

16. In a series of tests carried out with a loaded train of two-axled vehicles with wedge-loaded dampers, it was clear that the friction levels in the dampers were high enough to lock the suspension for a substantial proportion of a journey. During those periods, the unsprung mass was 25t per axle.

17. As part of the same test series the change of ground vibration level with speed was investigated, and was found to follow the curve sketched in Fig. 4. The peak at low train speed seems to be the speed at which the sleeper passing frequency corresponds to the vertical bounce frequency of the total vehicle mass on the track stiffness.

18. Other measurements were carried out at the side of a line which carried a service of 100t, 4-axle wagons loaded with dry pelletized material. Complaints from lineside residents had followed the introduction of this service. By arranging for the trains to pass the measurement site at a range of speeds, it was possible to examine the way in which the wayside vibration spectrum in the ground varied as the train speed changed. Figure 5 is a so-called Campbell diagram, and shows the way in which the position of peaks in the spectrum change as the train speed changes. Where the frequency of a peak is proportional to train speed, the plotted points lie on a series of sloping parallel straight lines, each corresponding to a particular "wavelength" of a repeating feature (eg wheel circumference, vehicle length, sleeper spacing, rail length, etc). Where the frequency is independent of speed (eg vertical bounce frequency of the vehicle), the peaks would lie on a horizontal line. The results suggest strongly that features of the track are important. Sleeper pitch (ca 700 mm) shows up clearly, and is almost invariably identifiable in wayside vibrations, though not usually at a nuisance level. Long wavelength corrugation

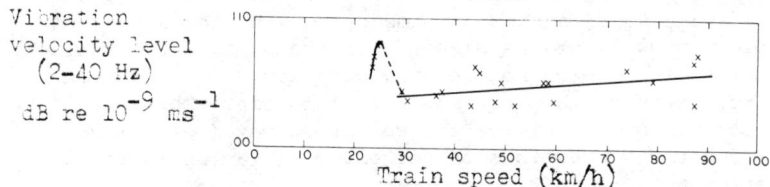

Vibration velocity level (2-40 Hz) dB re 10^{-9} ms^{-1}

Train speed (km/h)

Fig. 4. Ground vertical vibration level at 25 m from track

(Ref. 5)

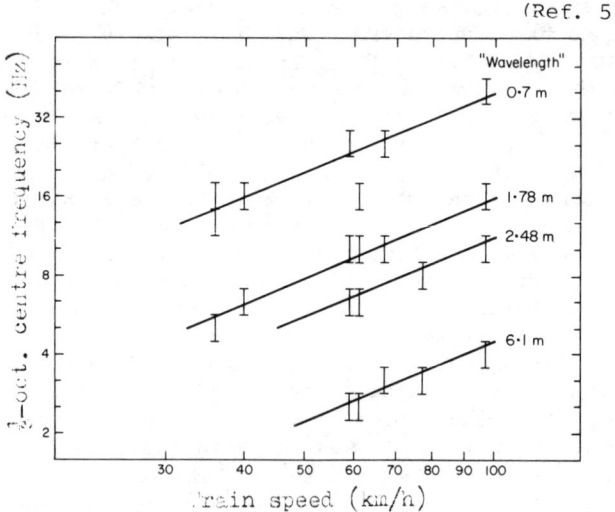

Fig. 5. Position of peaks in train vibration spectrum vs. train speed

(Ref. 6)

(ca 1780 mm) is also present, together with a wavelength around 2½ m. 6.1 m is one half of the vehicle length and one third of the rail length; every time the train advances 6.1 m a pair of bogies passes a rail joint.

19. The conclusions from these early investigations are:

- Not all vehicles create vibrations at a nuisance level, but high unsprung mass is an undesirable feature.
- Those which do cause vibration at a high enough level provoke vibrations characterised by features of the track.

20. More recent work seems to emphasise that track characteristics are important. A calculation based on a simple model of a coal train passing over a track whose longitudinal profile had been measured over a substantial length, and taking into account the elastic properties of the track and ground has shown an encouraging degree of agreement with vertical vibration levels measured at the wayside.

21. Much remains to be explained, however. Not only vertical vibration is caused - there are roughly equal amounts of vibration in the ground with horizontal particle motions parallel and perpendicular to the track, often with different spectra to each other and to the vertical component. The origin of some of this horizontal vibration is difficult to envisage.

22. Investigation has shown the type of ground over which the track is laid to have a large influence. Although seismic survey techniques have been used extensively to determine the elastic properties of the ground; it has not yet been possible to relate the vibration measured to specific ground properties. Generally, however, the softer the ground, the greater the vibration level.

BUILDING RESPONSE

23. Some measurements of the effects of passing trains have been made within and near a few houses. Whilst insufficient measurements have been done to produce a statistically reliable picture of the way houses in general can be expected to behave, dramatic effects can be expected in some instances. Figure 6 shows an example of the dynamic response of the floors in a two-storey house of traditional design. There has been a substantial increase in vibration level between the ground and the floors in the frequency range 20 to 40 Hz, but at higher frequencies the response is substantially reduced.

REDUCTION OF VIBRATION NUISANCE

24. It is sometimes suggested that reducing train speed will reduce wayside vibration, but this is not a very strong effect, unless the train speed is near some critical speed determined by the track and vehicle characteristics. In this case a change of speed in either direction might be beneficial (see Fig. 4), but train speed reduction could make things worse.

25. Some of the remedies lie with the Mechanical Engineer,

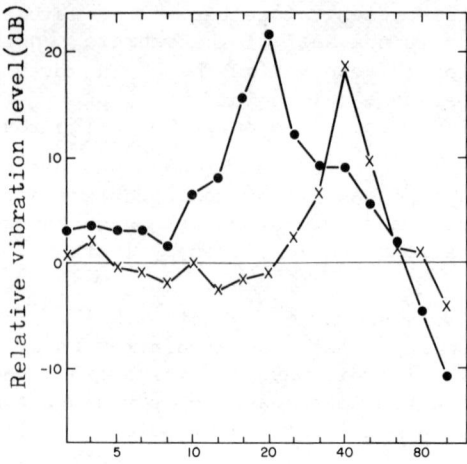

Fig. 6. Building vibration levels
relative to the ground
-●-, first floor; -x-, ground floor
(Ref. 6)

who should keep unsprung mass to a minimum, and make sure that
vehicle dampers are not so stiff as to "freeze" the suspension.
26. However, a substantial part of the problem lies with the
track. Although some vehicles are worse than others, the
wayside vibration is almost always determined in character
(ie spectrum) by the characteristics of the track and by the
speed of the train. If it is assumed that the problem is due
mostly to heavy freight trains, and these have service speeds
up to 100 km/h (28 m/s), it becomes possible to suggest what
"wavelengths" of track feature might be responsible. It has
been seen that railway vibration lies largely between 8 Hz
and 60 Hz, but this is also the frequency range which embraces
building resonances. The railway is therefore vulnerable to
influences quite outside its control, and it can only seek to
limit the energy generated at these frequencies.
27. The remedial action is to maintain the track so that
the oscillating component of force on the ground due to the
train is kept low. The geometrical consequence is that the
longitudinal rail profile should be kept as true as possible
for "wavelengths" between 3.5 m and sleeper pitch. Variations
of vertical dynamic stiffness should be made as small as
possible over the same wavelength range. The order of
improvement necessary should emerge more definitively from
the current lines of research.
28. If geometrical improvement is not possible in the worst
case (some heavy traffic on "soft" ground) then it is possible
that the alternative of incorporating additional resilience
in the track structure may have to be considered. There is
as yet no proven system for heavy traffic, but resilient
baseplates, sleeper soffit pads and under-ballast mats are all
being considered as modifications to classical track. Various

systems involving massive concrete slabs in the track structure are also under trial.

CONCLUSIONS

29. The reaction of people to the vibration produced by a railway seems to be described by an integration of vibration acceleration of the form:

$$\text{vibration dose} = \left(\int_0^T a^4(t) \, dt \right)^{1/4}$$

where a(t) is the frequency - weighted vibration acceleration time history. The weighting employed is the z-axis weighting of BS 6472.

30. Taking this together with the experimentally determined factors which are concerned in vibration generation suggests:

- The effects due to the worst trains should be given priority.
- The factor-of-improvement required is in the range 2 to 4.
- The feature "wavelengths" in the track which require attention are from 3.5 m down to sleeper pitch. Both longitudinal profile and vertical stiffness variations will be important.

ACKNOWLEDGEMENTS

31. The authors would like to thank their respective organisations for agreeing to the publication of this paper.

REFERENCES

1. British Standards Institution, Evaluation of humam exposure to vibration in buildings .(1 Hz to 80 Hz) BS 6472 (1984)
2. Internation Standards Organisation. Guide to the evaluation of human exposure to vibration and shock : Acceptable magnitudes of vibration. Draft Addendum ISO 2631/ DAD1 (1979).
3. Human Factors Research Unit, ISVR, Southampton University. Experimental studies of subject response to railway induced vibration. DT 159. Office de Recherche et d'Essais de l'Union International des Chemins de Fer, September 1983.
4. Fields J. M. and Walker J. G. The response to railway noise in residential areas in Great Britain. Journal of Sound and Vibration (1982) 85 177-255.
5. Dawn T. M. Ground vibrations from heavy freight trains. Journal of Sound and Vibrations (1983) 87 (2), 351-356.
6. Dawn T. M. and Stanworth C. G. Ground vibrations from passing trains. Journal of Sound and Vibration (1979) 66 (3) 355-362.

18

Noise and vibration reducing track foundation for subways and rapid transit railways

J. EISENMANN, L. STEINBEISSER and F. DEISCHL, Technische Universität München, Prüfamt für Bau von Landverkehrswegen, Federal Republic of Germany

1. INTRODUCTION

The reduction of noise and vibration in the vicinity of tunnel buildings has become more and more important during the last few years. This was on one hand caused by the rapidly growing number of subway lines in built up areas. On the other hand people are now realizing much more the whole problem of emission and pollution.

In the following we will try - without going too much into physical details - to survey the present state of research on noise and vibration and the development of means in order to reduce them.

2. THE EXCITATION AND JUDGEMENT OF NOISE AND VIBRATION

2.1 Excitation

Noise and vibrations are caused by discontinuities and irregularities of both the running wheel and the rail. If the tunnel building is excited by them to oscillations, elastic waves are sent out into the soil, which may force the foundations of buildings to oscillate, too. The critical frequency band extends to about 200 Hz. The so caused oscillations of buildings may have the following effects:

- very low resonances of buildings may be felt. Moreover they may cause malfunction and even destruction of delicate equipment.

- resonances of parts of buildings, which may act as a sound source and cause a so called secondary noise. The most critical range lies here between 40 and 80 Hz, coinciding with the natural frequency of walls and ceilings.

Damage of buildings or dynamical consolidation of the foundations soil, caused by the operation of underground railways occur very rarely.

2.2 Judgement

A regulation by law for the limits of noise and vibration, caused by underground railway in buildings, does not yet exist in Germany. So in practice some guide lines, set up by VDI (Comittee of German engineers) are used. For the judgement of noise the usual dB(A)- values are used, while low frequency vibrations are classified by their so called KB-value, with is derived from the velocity of vibration. These guide lines provide the following limits for night-time between 22^h and 6^h :

- Noise limit: 35 dB(A) peak value. This limit is of course much to high, if concert halls, churches a.s.o. are concerned.

- Vibration limit 0.1 - 0.2 (KB) depending on the site of a building (industrial site or residential district). Here the problem is that the human susceptibility against vibrations varies very much. There are persons with a sensation limit well below 0.005 KB (mm/s)

2.3 Present State of Research and Development of Insulating Measures

While vibrations can be measured very exactly, till now the transmission from one place to another is difficult to predict. In a research project, which is sponsored by German Ministry of Research and Technology since about four years, in almost every aspect investigations have been carried out. The results show that the most effective and economical measures in order to reduce vibrations are those performed on the permanent way. Here on every critical spot of a specific track special measures can be carried out, while measures on the rolling stock for example are in vain at less critical track sections. For the following, the understanding of two things is important:

- The vibrations level is normally given in the logarithmic dB scale

$$L_v = 20 * \log (v/v_0)$$

where v is the measured velocity of the vibration and the value

226

$$v_0 = 5 * 10^{-6} \quad cm/sec$$

which can be derived theoretically: If vibrations
of an oscillating wall and the air-borne noise
caused by this oscillation are measured at the
same time, this value has to be set in order to
obtain identical values for L_v and the unweighted
noise level.

It is of course a property of the dB scale, that
doubling the velocity leads to an increase of 6 dB.
For the judgement of vibrations, from the signals,
recorded at the measurement site, usually the mean
square values and the transformation from the time
domain into the frequency domain is calculated.
The result of many measurements, performed by German Railways and Prüfamt für Bau von Landverkehrswegen in subway tunnels is composed in Fig. 1. The
transducers were mounted rectangular to the tunnel
wall in a distance of about 1.5 m from the top of
the rail.

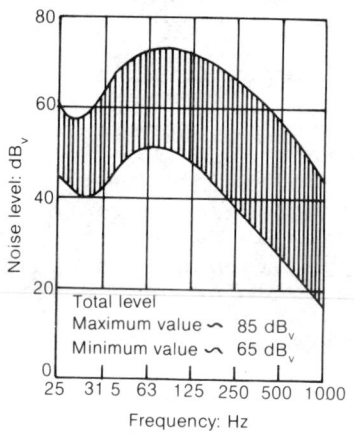

Fig. 1

The reduction of vibration can be achieved by damping or insulating. Damping means lessening the
energy of vibrations by transforming it into heat
(sub-ballast mats). Insulation takes place if,
for example, waves are reflected or if a
sinusoidal excitation with a certain frequency
strikes a linear oscillator with a natural frequency much lower then that of the excitation
(mass-spring system).
In practice, these two effects mostly occur commonly, with one of the two prevailing more or
less, so that one speaks summarily about the
damping effect.

227

We will now summarize the measures which can be
carried out on the permanent way in order to
reduce the emission levels.

- Rail surface

Short and long pitch corrugations, insulating
joints, turn outs a.s.o. are frequently the reason
for high noise levels. A corrugated rail for example may raise the noise level by as much as 10 dB.
The same increase occurs on switches. Generally a
good condition of the rail surface is important.
High strength rails may make maintenance less
troublesome. Switches, points and crossings could be
equipped with movable fogs. The great increase in
service life may compensate the higher prime costs.

- Damping measures on the rail web

Investigations have shown that covering the rail
web with damping material may lessen the noise
levels by about 2 - 5 dB. This means that normally the improvement achieved does not pay off.

- Highly elastic rail fastenings

Highly elastic fastenings permit a higher deflection of the rail under the wheel which reduces the
mechanical impedance of the superstructure which
in turn may lessen the excitation of vibrations.
The limits here are set by the fatigue strength of
the rail, the gauge widening and faulting in the
case of rail cracking. The damping effect here
begins at about 30 Hz and reaches the amount of
6 - 10 dB at about 50 Hz.

- Variation of the thickness of ballast

Normally ballast is about 30 cm thick. An increase
of ballast thickness has no measurable effect
while decreasing thickness below 30 cm leads to a
noticeable deterioration. There is also quite a
big difference between a recently tamped ballast
and a ballast that has not been tamped for a
long time, which again puts emphasis on the necessity of track maintenance.

- Sub-ballast mats

There is offered a great variety of products,
which also vary much in their effectiveness.
Problems can arise with respect to the durability,
which affects the track bed of course. Very good

mats lessen the vibrations at about 20 Hz noticeably and reduce them at 50 Hz by 20 dB. This
reduction may even be increased by about 4 dB, if
the thickness of ballast is increased from 30 cm
to 60 cm, a measure which is useless without
mats.

- Mass-spring system

This is normally the most effective, but as well
the most expensive measure. A lot of different
constructions have been built in the last few years.
The basic idea is to prevent vibrations from
penetrating into the soil by inserting a linear
harmonic oscillator with a very low natural
frequency between superstructure and foundation.

The natural frequency should be as low as possible in order to let the insulation of vibrations take place already at the low end of the
frequency spectrum.
Due to practical limitations, one cannot come
below 8 Hz but there is no necessity to exceed
14 Hz.

The mass of the oscillator is always a concrete
slab which may be prefabricated or built in site.
Care should be taken to dimension this slab correctly, in order to avoid low frequency resonances
of the slab itself.

On this slab the track is layed. Here an unbal-
lasted track seems to be the most suitable type of
superstructure, but it may cause noise problems in
the trains. In general the mass, which can be
brought in ranges between ca. 4 and 8 to/m, which
leads to a total height of ca. 65 to 120 cm for the
construction,- measured from top of rail to the
tunnel base.

The spring usually consists of rubber or plastic
elements. They should have a good linearity over
the whole range. If we neglect the damping of the
spring, a displacement of the slab of f (cm)
yields a natural frequency of:

$$n = \frac{5}{\sqrt{f}} \quad Hz$$

If we have, for example, a displacement of 2.5 mm
under the weight of slab and superstructure, a
natural frequency of 10 Hz results. In order to
avoid overloading of the rails and rubber elements, the additional displacement, caused by

trains should not exceed twice the displacement
resulting from the static preload.

3. RESEARCH WORK AND CONSTRUCTIONS PERFORMED BY THE PRÜFAMT

Nearly all types of vibration reducing tracks have
been investigated during the last few years.

Research work has been done on components like
sub-ballast mats as well as on existing tracks in
tunnels. In the following the two most effective
vibration reducing measures are described in more
detail.

3.1 Mass-Spring System

In 1968 till 1970 two systems have been developed at
Prüfamt by Eisenmann, one for the tram way in Zürich
and the other for tunnels of German Railways in
Munich and Frankfurt. Both system have been in
service since that time without any problems.
Comprehensive field measurements have been carried
out on both systems, which on one hand showed their
effectivity and on the other hand gave a number of
insights into the behaviour of the system.

In the meantime mass-spring systems have been
installed at critical spots of underground railways

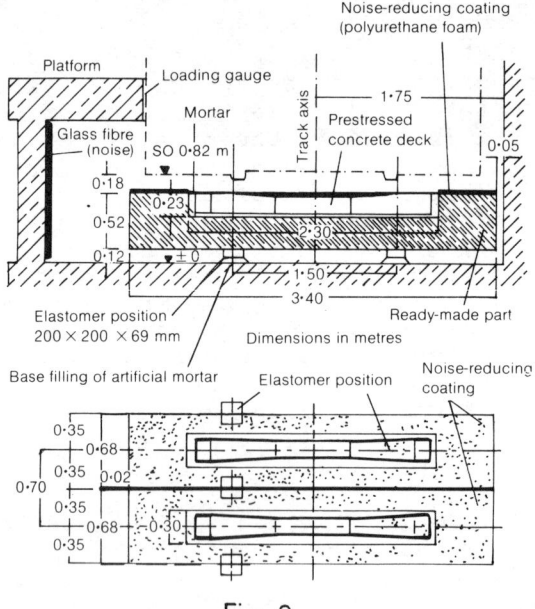

Fig. 2

in Dortmund, Düsseldorf, Hamburg, Ludwigshafen, Köln, München, Nürnberg and Stuttgart.

Fig. 2 shows a typical construction without ballast.

The prefabricated base slab is 70 cm wide and concrete sleepers are set in cement mortar.As spring elements elastomer bearings are employed like in bridge building. Fig. 3 shows the construction with

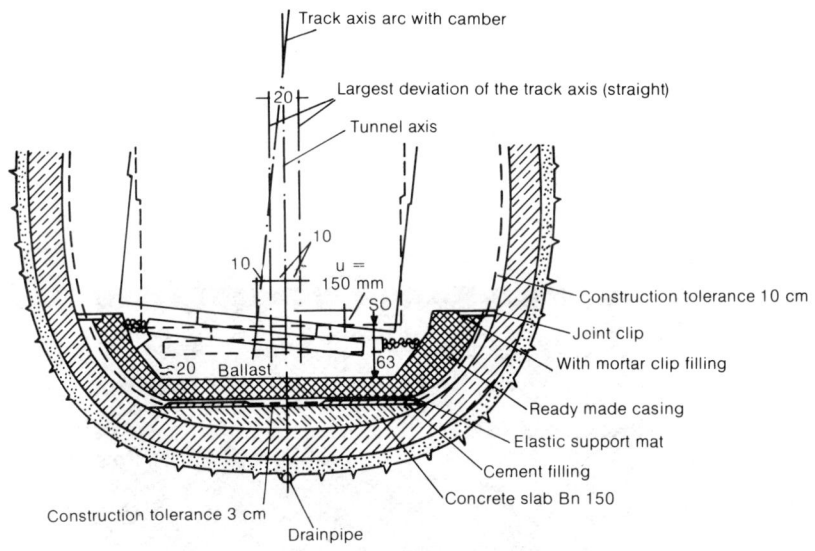

Fig. 3

ballasted track used at U-Bahn, Munich. The prefabricated concrete troughs are 100 cm wide; they are dowelled together in order to fix them in position. The weight is 5 to/m.

Any measure on the track must not endanger the operation, of course. Therefore every component used must be tested adequately. The reliability of a mass-spring system depends mainly on the bearings. They should not change their elastic behaviour. It is important that all lateral and vertical forces are transmitted through the bearings without any additional support. Bearings having been in service for more than 10 years at the S-Bahn were removed and checked. They still showed a perfect working.

231

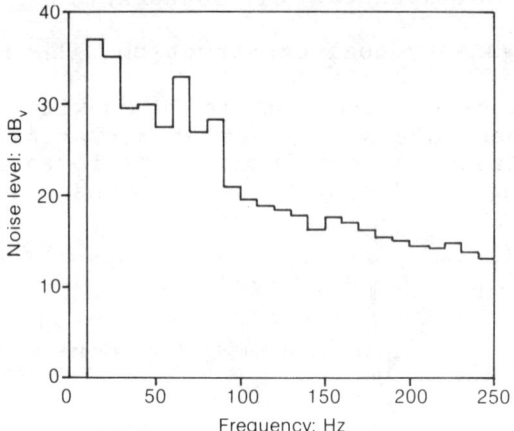

Fig. 4

Fig. 5

In Fig. 4 the results of a vibration measurement
performed at the wall of a tunnel, equipped with the
construction of fig. 3 are given. As one can see,
the vibration level is very low compared with that
of a normal ballasted track (Fig. 1).

In order to be able to precisely dimension a mass-
spring system, a test bed was developed, which is
capable of exciting the whole superstructure in
the frequency range from 10 to 70 Hz. In this way
all components of a mass-spring system can be tested
under realistic conditions (Fig. 5). The test bed
can be operated in the tunnel, too. Measurement
results gained here give a good basis for the
prediction of a possible vibration emission into
buildings.

3.2 Sub-ballast mats

The first sub-ballast mats were built in the early
seventies. The testing methods and the quality
standards were set up according to those early
prototypes. Measurements showed that there was much
room left for an improvement of the damping effect.
In consequence, a great variety of different
mats was developed. It was noticed that the
classification of mats by their dynamical stiffness
was insufficient, or even misleading. On the other
hand, a judgement of the damping effect of different
mats by field measurements was difficult, due to the
differences in the various measurement sites
(vehicles, soil, tunnel construction a.s.o.). Here
the test bed, mentioned above, proved to be a great
help.

In general, there are the following requirements, a
sub-ballast mat must meet:

- The damping effect should be high enough and must
  remain the same over the whole service life.
- The mat must not be destroyed by the operation of
  trains, nor must its softness have an influence
  on the adjustment of the track nor must it damage
  the crushed ballast.
- Building in a mat should not hamper the usual
  track maintenance.

4. CONCLUSION

The necessity for emission protection has caused
quite a lot of activity in the investigation of the
reasons and the measures to reduce noise and
vibration. It was found out that the most
satisfying and economic measures are those performed

233

on the track. For almost every vibration problem
there are nowadays effective means available, for
example mass-spring system or sub-ballast mats. For
underground railways with axle loads ranging from
120 to 210 kN meanwhile guide lines for the
application of protective measures have been set up.

In some cases, track sections, having produced a too
high emission level, were cured by the application
of sub-ballast mats. A subsequent installation of a
mass-spring systems has not yet been carried out,
but is under discussion for a track section of
S-Bahn in Hamburg.

REFERENCES

Eisenmann, J.:
Beanspruchung des Eisenbahnoberbaues und seine
Weiterentwicklung für höhere Geschwindigkeiten und
Achslasten.
ETR-Eisenbahntechnische Rundschau 17 (1968),
H. 5, S. 184-196

Eisenmann, J.:
Oberbau bei Stadtbahnen und U-Bahnen unter besonderer
Berücksichtigung der Korperschallemission.
I V-Internationales Verkehrswesen 33 (1981),
H. 1, S. 44-48

Steinbeißer, L.:
Körperschallmessungen in Zürich und München.
VDI-Berichr Nr. 273, 1974

Steinbeißer, L.:
Der gegenwärtige Stand der Körperschallmessungen und
Auswertetechnik im Eisenbahnoberbau.
AET - Archiv für Eisenbahnbechnik, Folge 30 (1975),
S. 33 - 40

Hölzl, G.:
Körperschall- bzw. Erschütterungsausbreitung an
Schienenverkehrswegen.
ETR - Eisenbahntechnische Rundschau 31 (1982),
H. 12, S. 881 - 887.

Braitsch, H.:
Das Kölner Ei (Oberbau 1403/c).
Verkehr und Technik 32 (1979), H. 7, S. 285 - 289,
H. 8, S. 323-238, H. 10, S. 460-465 und H. 12,
S. 523-530

Uderstüdt, D. und G. Eckermann:
Geräuschbelästigung durch unterirdische Verkehrs-
anlagen, Zeitschrift für Lärmbekämpfung 25 (1981),
H. 28, S. 8-19

Kohler, K.A.:
Polyurethan-Elastomer für die Körperschall-Dämmung
im Eisenbahnoberbau.
Verkehr und Technik 35 (1982) H. 2, S. 45-49

Eisenmann, J., Steinbeißer, L. und Deischl, F.:
Körperschallemission bei S-Bahnen in Tunnellage;
Großprüfstands-Messungen und Empfehlungen für
die Praxis.
ETR-Eisenbahntechnische Rundschau 32 (1983),
H. 12, S. 831-838

19 Subgrade requirements and the application of geotextiles

Professor G. P. RAYMOND, Department of Civil Engineering, Queen's University, Kingston, Canada

SYNOPSIS. A design method for calculating the stresses below the crossties of conventional railway tracks is presented. A method is outlined of their use to estimate the required ballast plus subballast depths on various subgrades based on the Casagrande classification for soils. Also mentioned is the use of geotextile permeable membranes for both track rehabilitation and new construction.

INTRODUCTION

1. The two main requirements of a stable subgrade are the provision of sufficient granular or soil modified cover to ensure that overstressing does not occur and the provision of a granular filter blanket to prevent the piping and thus loss of subgrade fines from below the track bearing area. In order to ensure that overstressing does not occur the track stresses need to be calculated. A means of calculating these stresses is outlined first. Discussion of subgrade stability techniques is presented later.

STATIC DESIGN OF CONTINUOUS WELDED RAIL

2. Talbot (Ref.1) · has shown that the theory of a continuous beam on an elastic support or, more correctly, a linear spring (Winkler) foundation gave calculated bending moments for the rails close to those measured in freshly tamped track. Many investigators have confirmed the validity of this theory.

3. This design method results in the representative differential equation for the elastic beam (rail) subject to a vertical (wheel) load P of

$$EI \frac{d^4z}{dx^4} + Uz = 0 \tag{1}$$

where E is the elastic modulus of the beam, I is the second moment of area of the beam, z is the deflection at a point x from the applied load, x is the distance along the beam from the point of application of the load, and U is the modulus of the elastic support commonly known as the track modulus.

Note the track modulus as used in common practice relates to a single rail. For a single wheel load P on an infinitely long beam the solutions to this equation are well known and are

$$z_x = \frac{\lambda P}{2U} \quad (\cos \lambda x + \sin \lambda x) \exp(-\lambda x) \qquad (2)$$

$$M_x = \frac{P}{4\lambda} \quad (\cos \lambda x - \sin \lambda x) \exp(-\lambda x) \qquad (3)$$

where $\lambda = [U/(4EI)]^{0.25}$ (4)
The maximum values of z and M are directly below the load at x = 0.

$$z_o = \lambda P/(2U) \qquad (5)$$
$$M_o = P/(4\lambda) \qquad (6)$$

These results are normally given in the form of an influence chart where the distance to zero moment, X_1 (given below), is used as a distance base

$$X_1 = \pi/(4\lambda) \qquad (7)$$

From the above, the rail seat loads are commonly calculated as

$$Q_x = z_x US \qquad (8)$$

where S is the crosstie spacing. For a crosstie placed directly below the load (x = 0)

$$Q_o = P\lambda S/2 \qquad (9)$$

4.   As an example of the effect of track modulus on rail seat loads Figure 1 shows the variation of the rail seat loads in the vicinity of a single 147 kN (33 kip) wheel load on 68 kg/m (136 lb/yd) rail for a 610 mm (24 in.) crosstie spacings in which the central crosstie (crosstie No 1) is directly below the load (i.e. the crosstie is positioned to produce a maximum rail seat load). Examination of Figure 1 and Equation (9) indicates that a variation of the track modulus has a much greater influence on the rail seat load than does the crosstie spacing.

EFFECT OF MULTIPLE WHEELS
5.   Since the above theory is based on elastic response the theorm of superposition applies and the effect of multiple wheels or axles can be simply calculated. Results from such calculations showed that as the track modulus decreases from between 35 to 70 MN/m/m (5-10 kip/in/in) there was a rapid increase in deflection or rail stress while if the track modulus increases from these values a gradual reduction of deflection or rail stresses occurred. It was thus concluded that for statically loaded track 35 to 70 MN/m/m (5-10 kip/in/in) may be anticipated as an

optimised value range of track modulus, at least for initial consideration.

CALCULATIONS OF STATIC SUBGRADE STRESSES

6.   Similar calculations can be made for the determination of the rail seat loads from multiple axles and thus tie bearing pressures through to subgrade stresses.   Sufficient evidence is available in the technical literature to show that where soils are unable to resist tensile forces the vertical stress is given by superposition of Boussinesq's solution for stresses within a semi-infinite elastic solid with surface loading.   Below each rail seat the crosstie can be assumed to produce a rectangularly loaded area to the ballast.   This permits the use of the solutions developed by Love (Ref.2) who extended the Boussinesq's solution to a rectangularly loaded area.   The area of the tie bearing surface per rail seat, A_b, may be calculated as equal to some tamper influence distance either side of the rail.   For purposes of illustration rail seat bearing lengths of 914 mm (36 inch) below two rails spaced 1.50 m (60 in) centreline to centreline apart will be assumed.

7.   Figure 2 shows the solution obtained using different track moduli of the underlying vertical support stress resulting from a single axle load on 279 mm (11 inch) crossties spaced at 610 mm (24 inch) centres.   The contact pressure or vertical stress at zero depth may be seen to increase as the fourth-root of the track modulus (i.e. an increase of track modulus by a factor of sixteen doubles the contact stress).   With depth the difference in vertical stress decreases slowly such that, even at 0.8 metres (32 inches) the effect of track modulus variation on subgrade stresses from the single axle are considerable.

8.   Figure 3 shows similar calculations, using the same crosstie size and spacing, for a 3 axle SD locomotive truck. Interaction between axle loads is sufficient under the lower values of track modulus to cause a considerable increase in the contact pressure or vertical stress at zero depth.   This partially reduces the dependence of either maximum contact pressure or maximum subgrade stress on the variation of track modulus.   Using the two highest track moduli interaction actually caused a reduction in these maximum values from those calculated for a single axle.

9.   It is clearly evident from these results that stiffer track results in higher support stresses through the ballast and into the subgrade.   In the case of a single axle on 68 kg/m (136 lb/yd) rail changing the modulus from 14 MN/m/m (2000 lb/in/in) of rail to 224 MN/m/m (32,000 lb/in/in) of rail causes the crosstie/ballast interface stress, for the same crosstie spacing, to double (i.e. 100% increase). Interaction of axles is more pronounced on softer track since the rail is more effective in spreading the load. Axle load interaction for the 3 axle SD locomotive truck, for the same rail and crosstie conditions, reduces the interface stresses from the above modulus change alone to an increase of approximately 40%.

239

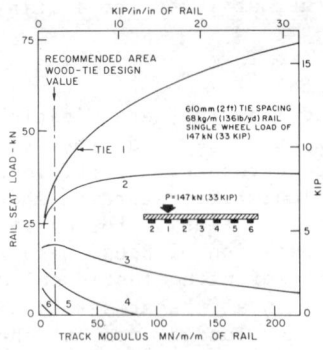

Fig.1. Rail seat loads

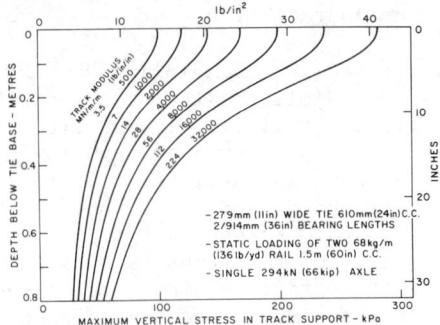

Fig.2. Stresses below single axle

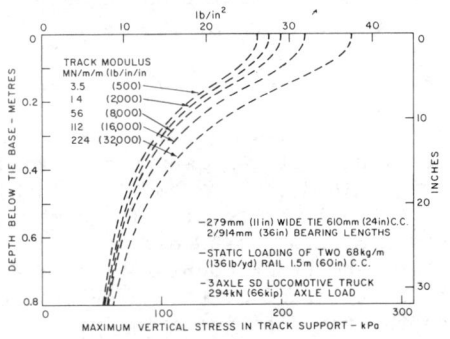

Fig.3. Stresses below SD
locomotive truck

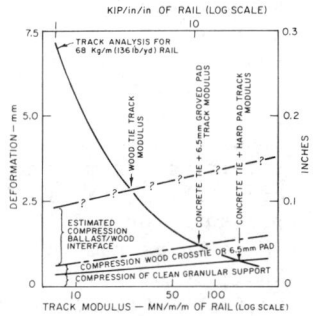

Fig.4. Estimation of
track modulus

Fig.5. Placement of geotextile on capping sand

ESTIMATE OF TRACK MODULUS

10. The solutions developed by Love (Ref.2) for a rectangular loaded area allow for the calculation of not only the vertical stress but rather the six stresses (3 normal and 3 shear) that define the state of stress at any point. Knowing the pseudo-elastic parameters of the support soil allows the calculation of the strain at that point which, by numerical integration, allows the calculation of support deformation. Addition of such deformations to those produced by the rail seat load on the crossties via the tie plates allows a total deformation to be estimated for a given set of conditions. Since different track moduli produce different calculated support stresses and rail seat loads a series of calculations of total deformations are required using different track moduli. These results should then be compared with the deformations predicted by the 'Beam on Elastic Support' theory to determine the track modulus which defines identical deformations. It should be clearly understood that the deformation experienced by the rail at the rail seat includes not only the deformation of the track support but also the crosstie and its accessories. An example of the results of such a series of calculations is shown in Figure 4. These results have been done for a single axle loading since many field measurements of track modulus are obtained using a long loaded flatcar with central loading jacks. The jacks are used to apply in increments the equivalent of a single axle load of different magnitudes.

11. The results given in Figure 4 were obtained using 305 mm (12 in) of clean ballast overlying 610 mm (24 in) of clean subballast overlying a compacted silt or clay subgrade. It may be seen that track on such a support constructed with concrete crossties and stiff pads would have a track modulus of about 14 MN/m/m (20,000 lb/in/in) of rail. Use of the softer grooved pads recently installed on the U.S. Northeast Corridor reduces the track modulus to about 6.9 MN/m/m (10,000 lb/in/in) of rail. The compressibility of hardwood crossties is similar to the Northeast Corridor Pads however the major compression observed in tests on wood crosstie track has been the elastic compression/rebound of the ballast penetration of wood or ballast/crosstie interface. The magnitude of this interface compression is shown on Figure 4 from which it is seen that wood crosstie track has a considerably lower track modulus than concrete crosstie track. The curves plotted on Figure 4 suggest a track modulus of 2.4 MN/m/m (3500 lb/in/in) of rail, however the variability of the interface compression means considerable variation from this value would not be unrealistic.

CALCULATION OF DYNAMIC INCREMENT FOR CONTINUOUS WELDED RAIL

12. Kenney (Ref.3) has shown that on perfect track transversed by perfect wheels track forces rise very slowly with speed requiring a speed of about 1600 km/h (1000 mph)

to reach track resonance. Track irregularities or wheel irregularities are therefore the principal cause of major dynamic track forces based on present day (1984) speeds. Any complete theory dealing with track/wheel irregularities is clearly complex but since the track mass is much less than the unsprung mass of a wheelset it seems reasonable to neglect the track mass as a first approximation and to take the effect of the suspension spring as a steady force. The equation of motion then reduces to:

$$M \frac{dw^2}{dt^2} + K (w + s) = 0 \qquad\qquad (10)$$

where w is the displacement of the unsprung mass from the static equilibriunm position, s is the amplitude of the irregularity, M is the unsprung mass, and K is the track stiffness. The track stiffness being related to the track modulus by

$$K = (64 \, E \, I \, U^3)^{0.25} \qquad\qquad (11)$$

The solution to equation (14) leads to a relationship for dynamic increment of load of the form

$$P_d - P_o = C \, s \, (M \, K)^{0.5} \, f(V) \qquad\qquad (12)$$

where P_d is the dynamic wheel load, P_o = the static wheel load, C is a proportionality parameter, and f(V) is a function of speed.

13. For a well maintained track made of continuous welded rail the major track/wheel irregularity is due to wheel flats. Data on wheel flat impact obtained by the Association of American Railroads has been reported in Ref.4. According to Equation (12) for any given speed and weight of rail the dynamic increment should increase linear with irregularity. Since the wheel flats were carefully made as square flats the increase in irregularity with flat size may be calculated and is considered to concur with the data. Also from Equation (12) the axle load should make no difference to the dynamic increment which is also considered in reasonable agreement with the data. Thus the dynamic increment from a given wheel flat for any given speed would reasonably be obtained from Ref.4 modified by the relationship of Equation (12) or

$$P_d - P_o \text{ is proportional to } M^{0.5} \, U^{0.375} \qquad (13)$$

The data of Ref.4 was obtained using a 70 (net) ton freight car having an unsprung axle mass of 1705 kg (3750 lb) on wood crosstie track. 100 (net) ton cars are in more common use today (1984) and these have an unsprung axle mass of about 1977 kg (4350 lb) representing an 8% increase in impact loading on wood crossties. If concrete crosstie

242

track with stiff pads is used this would represent an additional 98% while soft pads would reduce this to 55%. For both car and track changes the increment would be 105% and 59% respectively for the conditions represented in Figure 4. In fact, due to the compression of the components of the track structure from the static load increment the stiffness or track modulus associated with the dynamic load increment is probably not as numerically diverse as shown in Figure 4.

14. Based on the data presented, the dynamic wheel increment from a 51 mm (2 inch) wheel flat would be 197 kN (33000 lb) for wood crosstie tracks. If wheel flats were worn onto the same location of both axle wheels this increment would apply to each rail suggesting a dynamic axle load of 294 kN (66,000 lb). For the concrete crosstie tracks the axle increment would theoretically be 926 kN (208,000 lb) and 447 kN (100,000 lb) for stiff and soft crosstie pads respectively.

CALCULATION OF SUBGRADE STRESSES

15. The calculation of static stresses in the track support has already been outlined and plotted in Figures 2 and 3 for a single axle and a 3 axle SD locomotive truck respectively. A similar plot to Figure 3 may be done for any system of wheel loading such as two coupled trucks. To this static value the dynamic increment may be added from the distribution given from a single axle suitably proportioned to reflect the dynamic increment. For concrete crosstie track this dynamic increment of stress can be considerably greater than the static stress increment.

16. Once the maximum stress on the subgrade has been estimated, if necessary using smaller or larger wheel flats than used here as appropriate to the problem being analysed, and an appropriate factor of safety has been applied the values may be compared with the well known limits as given in Ref.5 and supplemented if necessary by in situ field tests. This will produce the depth of granular or soil modified fill required.

CAUSES OF UNSTABLE SUBGRADES

17. Irrespective of the amount of compaction a subgrade receives during construction some degree of permanent deformation occurs due to traffic loading. This loading is greatest below the rail causing the formation of a trench in which water may collect. On good subgrades this trench may only be millimeters deep but this is sufficient to cause water ponding and softening of the subgrade. To prevent silt size fines from penetrating upwards from any compacted fill or subgrade from below non-plastic fine sand sized material grading down to the No. 200 sieve needs to be included in the subballast or a non-plastic fine grained granular capping material needs to be provided between the fill or subgrade presently in contact with the subballast and the subballast itself. This capping material should act

as a filter to prevent fines from being vibrated and/or pumped upwards. Similarly subballast should be graded to prevent it from vibrating upwards into the ballast. A non-woven geotextile may be used to achieve this latter requirement as seen being done in Figure 5 but not to act as a filter to silts or clay. It should be clearly understood that the dynamic loading experienced within a railway track support fill is much greater than that generally experienced by any highway support system. As has been shown impact loading from wheel flats can impose dynamic loads several times those due to static loading. Thus the use of modified clay subgrades which form a brittle hard subgrade surface is generally subject to cracking and must be covered by a non-plastic granular filter material.

18. Thus even where subgrades or fills adjoining the subballast are well compacted summer dry weather may be expected to cause drying of the surfaces in contact with the subballast. Wetting after such drying weather may then be expected to cause collapse of the soil structure of the surface permitting the erosion of fines upwards into the open subballast. Over time such erosion could be expected to not only foul the subballast but also foul the ballast. Once fouled both materials may be expected to heave during freezing weather. The writer cannot overstress the importance of suitably grading subballast material such that it would remain unfouled by eroding fines from any underlying material and also be suitably graded so as not to itself be vibrated upwards fouling the ballast. This material must be non-plastic so as to deform easily and not permit vertical fractures.

19. The use of geotextiles for track rehabilitation has been dealt with in great detail in Ref.6. Where excavations are of insufficient depth to allow a capping sand below the geotextile a resin treated 1000 g/m^2 (29 oz/yd^2) non-woven polyester fiber geotextile has been recommended placed 300 mm (12 inch) below the crosstie base.

REFERENCES

1. TALBOT, A.N. Reprints of the Talbot Reports, American Railway Engineering Association, Washington D.C., 1980.
2. LOVE, A.E.H. The stress produced in a semi-infinite body by pressure on part of the boundary, Philosophical Transactions of the Royal Society, Series A, Vol.228, 1928.
3. KENNEY, L.T. Steady state vibrations of beam on elastic foundation for moving load, Journal of Applied Mechanics, Vol.76, 1954.
4. Association of American Railroads test data on wheel flats, American Railway Engineering Association, Bulletin Vol.53, 1952.
5. Portland Cement Association, Soil Primer, Chicago, 1956.
6. RAYMOND, G.P. Geotextiles for heavy haul railways, Canadian Geotechnical Journal, Vol. , 1984.

20   Ballast for heavy duty track

Professor E. T. SELIG, BSME, MSMech, PhD, Professor of Civil Engineering, University of Massachusetts, USA

SYNOPSIS. The mechanical performance of ballast in heavy duty track is reviewed from the perspective of US researchers. Topics include ballast requirements, fouled and cemented ballast conditions, compaction effects, and loading requirements. Also discussed are particle degradation, settlement and residual stress induced by repeated loading. Efforts to predict ballast performance in track are mentioned. Finally, some conclusions are given about the ballast conditions for best track performance, and some areas of needed future ballast research are indicated.

INTRODUCTION

1.   Ballast is an important component of conventional track structures. Its importance has grown with increasing axle loads and train speeds. Yet much is still unknown about the factors which affect the performance of ballast. Thus the choice of ballast type, gradation, and layer thickness is usually governed by tradition, availability, or oversimplified guidelines that do not integrate all of the important factors. What is needed is a basis for quantitatively ranking ballast conditions in terms of performance in track so that alternative ballast choices may be evaluated for optimum economy and safety.

2.   To address this need this paper first states the main requirements of ballast and the choice of material typically used to meet these requirements. In service ballast does not remain at its initial gradation because the voids become filled with smaller particles. Problems and questions associated with this fouled ballast condition are discussed.

3.   The compaction state of ballast in track affects track settlement. Thus the relationship of traffic and maintenance to compaction state based on field measurements of ballast density is described. Because traffic loading is closely associated with ballast behavior the characteristics of this loading are noted.

4.   A laboratory test for simulating the mechanical effects of traffic on ballast performance has been developed. The results of tests will be summarized to show the comparative degradation of ballast as a function of such factors as gradation, type of ballast, load level, and maintenance disturbance. The

significance of residual lateral stresses induced by repeated
load is also discussed as part of a brief commentary on track
performance prediction.

5. The paper concludes with comments on the implications of
past ballast research for improving track performance, and with
some suggestions for further ballast research.

BALLAST REQUIREMENTS

6. The most important ballast functions are to: 1) re-
strain ties against forces from train and track, 2) reduce
stress on the weaker subgrade, 3) facilitate maintenance, 4)
provide immediate drainage of water from the track structure,
and 5) provide resiliency to the track.

7. To achieve these functions the traditional choice of ballast is a gravel-sized crushed rock of relatively uniform gradation. In service, traffic load and the environment act to
convert the original gradation into a broader gradation. The
mechanisms involved include: 1) mechanically induced breakage, 2) mechanical wear, 3) chemical decomposition, 4)
breakage from freeze-thaw cycles, and 5) infiltration of fines
from surface and subgrade.

8. To minimize these effects, other than infiltration, a
hard, durable rock with particle sizes generally in the range
of 3/4 to 2 1/2 in. (19 to 64 mm) is considered best. The durability characteristics are assessed by such index tests as
abrasion resistance, sulfate soundness, crushing value, and
hardness. Each railroad has a set of ballast specifications
that stipulates limits for the values from the index tests.
These specifications are known to be inadequate for ensuring
satisfactory performance. One major limitation is that no correlation exists between index tests so that tradeoffs can be
established when a material is strong in one index test, but
marginal in another.

FOULED BALLAST

9. As the voids in the original, clean ballast become filled
with particles primarily smaller than the #200 sieve (0.075-mm
diameter), the ballast performance will be adversely affected.
This means that the ability of ballast to carry out stated
functions will diminish. In the advanced stages of this process of void filling the ballast is said to be fouled. Typically, for AREA ballast gradations, the porosity is on the order
of 35 to 50%. Thus fouling probably does not start to become
significant until the amount of fines becomes 10% or more.
However there are no generally established guidelines for this
condition.

10. Most commonly observed fouling problems are restrictions
of drainage and interference with track maintenance. However
as the voids become completely filled with fines the ballast
begins to take on the characteristics of the fines with the
ballast particles acting as filler. Soaked fines represent mud
and hence the ballast becomes soft and deformable. When wet
fouled ballast becomes frozen the resiliency is lost. When the
fines become dried (but still moist) they act as a stiff bind-

ing agent for the crushed rock particles. This also causes loss of resiliency. All of these conditions prevent proper track surfacing.

11. The term "cemented ballast" is frequently used in the railroad industry to represent a condition in which the ballast particles are bound together. Although this term has not been officially defined, in most cases it appears to be used to represent dried fouled ballast. However the word "cemented" has led to the notion that a chemical bonding is involved such as in the case of portland cement, a derivative of limestone rock. This is one of the reasons given by the railroad industry for preferring not to use limestone ballast.

12. A thorough examination of "cemented ballast" conditions is needed to determine the cause. Such a study could very well show that chemical bonding as in cement is not the main bonding mechanism in "cemented ballast" because it is not normally the type of bonding in dried fouled ballast.

13. Although rarely performed and not yet standardized, tests have been developed to measure the cement strength of fines obtained from ballast rock. Typically in this test a saturated paste of water and fines generated by abrading the ballast particles is placed in a small cylindrical mold. The paste is then compressed under a pressure of as high as 1875 psi (12,900 kPa). After the compressed cylinder has been oven dried its unconfined compression strength is measured. This strength is the cementing value of the ballast.

14. For cementing value to be a relevant measure of relative ballast quality, conditions in track must first cause the ballast to disintegrate into enough fines to foul the ballast. Since the porosity of new ballast is around 40%, this much degradation would cause a dramatic and easily noticed change in size and shape of the ballast particles. Then these fines must be compacted and dried in the voids presumably in a manner resulting in a state like that achieved by consolidating saturated ballast fines under high pressure as in the lab test. Such a combination of conditions would seem to be unlikely.

15. Field investigations are needed to establish the frequency of occurrence and nature of "cemented" ballast. Of particular importance is the determination of whether or not the fines in the ballast were derived from breakdown of the ballast, and, if they were, how. Only then can the relevancy of the cementing value test be established.

BALLAST COMPACTION

16. At the time when surfacing is required to correct track geometry irregularities, the ballast is in a dense state, particularly beneath the tie bearing areas. When the rail and tie are raised to the desired elevation, tamping tines are inserted in the crib next to the rail to displace the ballast into the voids under the tie that were created by the raise. This tamping process disturbs the compact state of the ballast and leaves it loosened. The more fouled the ballast is, and the greater the raise, the looser the ballast is after tamping.

17. The loosened ballast beneath the tie results in renewed settlement as the traffic recompacts the ballast. The loosened crib ballast results in a significant reduction in lateral buckling resistance of the rail in the unloaded state. Crib surface vibratory compactors can be used to redensify the crib ballast immediately after tamping, but not the ballast under the tie.

18. Traffic is the most effective means of compacting ballast under the tie, but this takes time and results in nonuniform settlement. Traffic also causes crib ballast to densify, although the reasons for this are not certain.

19. In addition to increasing density there is evidence that both the traffic and the crib compactor produce residual horizontal stresses in the ballast. These residual stresses may be one of the most important factors influencing ballast performance in track. Fouled ballast in the crib will reduce densification of crib ballast by traffic after tamping and hence diminish any tendency for the development of lateral residual stress against the sides of the ties.

BALLAST LOADING

20. Heavy duty track in the US is characterized as mainline track with tonnage mainly from freight service, but with passenger service added. Annually 25 to 60 million gross tons (223 - 534 GN) are typically carried on such tracks. Nominal wheel loads, defined as the static average assuming equal distribution to all wheels on a car, are 8 to 15 tons (71 - 134 kN). The lower end reflects a significant amount of passenger service, while the upper range is representative of loaded freight cars. Assuming continuously welded rails and reasonable equipment and rail maintenance, the average dynamic load has been found to be only up to 10 - 20% higher. However this average is misleading because the dynamic effects are very significant. For example 1 out of 1,000 wheels can apply a load of 3 times the static mean. Although these high loads were relatively infrequent, they contained high frequency components, well over 100 Hz, which compounds their damaging effect.

21. The computer model GEOTRACK was used to estimate the range of tie pressure applied to the ballast surface for a wide range of track conditions including tie stiffness, rail stiffness, ballast depth and tie spacing (ref. 1). For a wheel load of 16.5 tons (147 kN) the nominal ballast pressure was 38 psi (262 kPa) and the range was 17 to 65 psi (117 to 448 kPa). Higher pressures are expected when the ties are bearing primarily in the tamped area.

BALLAST EVALUATION

22. At present no adequate correlation exists between ballast index tests and ballast performance in track. What is needed to select ballast is a method that takes into account the effect of differences in ballast gradation and particle composition and, in addition, simulates field service conditions such as ballast depth, subgrade characteristics, traffic loading and track parameters. A start on this ambitious task has been

taken by a combination of analytical modeling and laboratory
box tests at the University of Massachusetts.

23. The box tests provide a measure of ballast performance in
terms of residual stresses, settlement, and particle degradation under mechanical conditions simulating the field situation
(ref. 2). The box contains a ballast section 12 in. deep by
12 in. wide by 24 in. long (30 x 30 x 61 cm). The bottom of
the box is a resilient support. An 11-in.-long (28-cm) tie
segment with a 7-in. (18-cm) height and 9-in. (23-cm) width is
placed in the center of the box with a 6-in. (15-cm) crib
depth. Vertical loading is provided by a servo-hydraulic actuator. The tie segment may be either wood or concrete.

24. This apparatus has been used to investigate the effect on
ballast performance of ballast type, ballast gradation, load
level, number of load repetitions, maintenance disturbance, and
mix of wheel loads. The effect of environmental conditions has
not yet been included.

25. Ballast density tests showed that increases in density
could be achieved under repeated loading in the box that are
comparable to those in the field under a tie caused by train
traffic. Plate load tests also showed increases in ballast
stiffness comparable to those measured in track. Tie settlement increased generally with the logarithm of the number of
load cycles. High residual horizontal stresses were observed
under the tie when the load was removed, with the biggest increase occurring in the first cycle.

26. In most of the tests the magnitude of repeated load was
held constant for all of the cycles. However for some tests
equal numbers of cycles for each of three load levels were applied in sequence. Regardless of the order of load application
in these multi-staged tests, the residual stress and tie settlement at the end of the test were almost entirely governed by
the highest load applied.

27. For the same ballast material, the ballast degradation
and settlement increased as the average particle size increased. Degradation also increased, as expected, with increasing wheel load, but the increase was much greater than
just proportional to load increase. Most of the degradation,
particularly the coarse breakage, occurred in the top 6 in.
(15-cm) of ballast beneath the tie. The crib zones by comparison had negligible degradation.

28. Maintenance, simulated by rearranging the top 6 in. (15-
cm) of ballast each 100,000 cycles, resulted in greatly increased ballast degradation, in terms of both coarse breakage
and fines generation.

29. For identical test conditions, except for ballast composition, ballast degradation and settlement correlated well with
the Los Angeles (LA) abrasion number. However when gradation
and particle shape varied significantly the LA values were less
meaningful. This situation is reasonable in part because the
LA test uses a standard gradation and so can not take into account the effect of gradation, while the box test results are
influenced by gradation.

30. The maximum amount of fine particles generated in the tests was about 1.6% of ballast weight, but typically was much less than 1%. This study suggests that mechanical wear of normal ballasts is insufficient to cause fouling. If so, fouling either must be a result of fines infiltrating from either the surface or subgrade. Chemical breakdown of the ballast is another potential mechanism for ballast degradation, but it has not yet been shown to be a major factor. A field investigation to confirm or disprove these possibilities is needed.

TRACK PERFORMANCE PREDICTION

31. A method has been developed to predict elastic response and permanent deformation of track with emphasis on the ballast and subgrade characteristics. The approach uses the computer model GEOTRACK (ref. 3) to determine stress states in the ballast. Elastic moduli and cumulative permanent strains are estimated from these stresses using measured ballast stress-strain properties.

32. The results of this work as well as field and laboratory observations strongly suggest that horizontal residual stress in ballast is a key factor that controls the accumulation of permanent deformation in ballast under repeated load. Of course, other factors, such as subgrade, affect track settlement, but the research indicates that any factors that influence residual stress in ballast will influence track performance. Such factors are traffic vibrations, insufficient crib and shoulder material, ballast degradation under load, and maintenance tamping.

33. Understanding of the role of the residual stress and means to represent it appears to be a key factor in improving methods of predicting permanent settlement of track caused by repeated train loading.

CONCLUSIONS

34. Recent research together with past experience suggests that the ideal ballast should consist of particle sizes distributed uniformly over the range from 3/8 in. to 2 in. (1 to 5 cm), with some particles beyond this range. The larger sizes are needed for stability and the smaller sizes for reducing the contact forces between particles. In addition the individual particles should be equidimensional rather than flat.

35. While this is a commonly used size range, the traditionally favored distributions within this range are generally much more narrow or uniform than what appears to be ideal from material performance considerations. A more uniform sized ballast will have a high porosity and so can accommodate more fines before performance is degraded by fouling. However the suggested gradation will tend to collect fines less easily and should perform better when not fouled. The main disadvantage of a wider gradation is that size segregation in handling and maintenance will be a more serious problem.

36. Ballast particles should also be durable under both mechanical and environmental influences. A wider particle gradation range as well as less frequent tamping will greatly re-

duce the mechanical degradation of ballast. Tests for evaluating freeze-thaw induced degradation of ballast are available. Otherwise environmental factors have not been given much attention. Chemically induced breakdown may be a bigger problem than mechanical factors and thus need to be studied.

37. Presently traffic provides most of the compaction of ballast. This is undesirable. Thus construction methods are desired that will place ballast in track without segregation and compact it equivalent to the effects of traffic while maintaining control of surface and alignment. Once placed and compacted, ballast beneath the tie should be left undisturbed until it needs to be cleaned. Thus, ideally, methods are needed for making adjustments in surface without conventional tamping. The British stone blowing process is a potential solution to this problem. Alternatively, if tamping is used, methods are needed for restoring the premaintenance density state of the ballast.

38. Effort must be made to insure that ballast does not become fouled. This requires, as a minimum, that protection must be provided against subgrade infiltration by a proper filter zone, which can be provided either by a suitably graded subballast, or a geotextile.

39. Proper subgrade conditions are needed before the optimization of ballast conditions becomes meaningful. Thus the ballast and subgrade must be adequately drained to prevent subgrade softening, erosion and pumping from trapped water. When poor drainage exists, most of the other considerations of ballast selection are of secondary importance. In addition to drainage the subgrade must be unsaturated and remain that way under the action of repeated wheel load.

40. Ballast layer thickness is an important track parameter that can be varied to influence track performance. Normally a 12-in. (30-cm) thickness is used for US heavy-duty track, but the suitability of this will depend heavily on subgrade conditions as well as on the service requirements of the track. At present in the US the choice of ballast thickness appears to be based on tradition rather than on track analysis. Better thickness guidelines are needed for defining ballast cleaning and reballasting requirements as well as for design of the occasional new or replacement track.

41. Assuming proper drainage and suitable subgrade conditions, refinement of ballast evaluation methods becomes worthwhile. Correlations between the traditional index tests and performance is too limiting and is not likely to achieve the desired goal. Instead further development of an integrated approach such as indicated in this paper is recommended. This involves a track model to relate track and traffic conditions to ballast loading requirements. The model provides input to a box test or some other property test that can measure the ballast behavior under expected field conditions. The ballast parameters such as gradation, composition, and particle shape must be represented. The scope of these property tests needs to be expanded to cover important environmental factors and vibratory

loading.

42. With the aid of the track computer model the ballast tests results provide a basis to predict the effect of ballast variables on track performance. Available models emphasize vertical loading. Models for lateral loading are also needed.

43. One of the most important steps that could be taken to advance the understanding of track performance as related to the role of ballast and subgrade is to conduct a well-planned, comprehensive field observation program. The purpose of this program would be to correlate track performance with ballast and subgrade conditions. Characteristics common to good and bad track performance need to be identified. Questions to be addressed include: 1.) when is ballast the principal factor in poor performance, 2.) under what conditions is subgrade responsible for poor performance, 3.) how does ballast become fouled, 4.) what is the nature of "cemented" ballast, 5.) how common is cemented ballast, and 6.) under what conditions is significant ballast degradation observed?

44. The success of this observation program depends on blending the knowledge gained through research with a careful site inspection and historical performance data for the track. The data should include: 1.) characteristics of the ballast and subgrade materials and a description of the track structure, 2.) maintenance history, including type of maintenance performed, frequency of maintenance and condition of track before maintenance, 3.) description of the general environmental conditions, 4.) traffic history in terms of annual tonnage and approximate mix of individual car weights, and 5.) performance experience. The railroads should establish a data management system for this information and routinely file the data for later retrieval and evaluation. Such an approach will help to identify and quantify the cause-effect relationships of track performance and replace the frequently used qualitative and subjective methods which generally are not based on adequate information.

REFERENCES

1. STEWART H.E. and SELIG E.T. Predicted and measured resilient response of track. Journal, American Society of Civil Engineers, Vol. 108, No. GT11, November 1982, 1423-1442.

2. NORMAN G.M. and SELIG E.T. Ballast performance evaluation with box tests. Bulletin 692, Vol. 84, American Railway Engineering Association, May 1983, 207-239.

3. STEWART H.E. and SELIG E.T. Prediction of track settlement under traffic loading. Proceedings, Second International Heavy Haul Railway Conference, Colorado Springs, Colorado, September 1982, 488-496.

21 Ballast deformation and track deterioration

Dr. M. J. SHENTON, Research & Development Division, British Railways Board, Derby, UK

SYNOPSIS. The deterioration of the vertical geometrical quality of track has been studied and computer programs developed for evaluating the different mechanism of deterioration. The main controlling factor is the settlement of the ballast, and the various parameters which influence this settlement have been investigated both in the laboratory and on the track.

INTRODUCTION

1. Every year all the railways of the developed world expend considerable effort and expense to maintain the geometry of the track to a level which will give both a safe and a comfortable ride. The components of the permanent way have been sufficiently developed that they normally keep dimensional integrity thus leaving the sole cause of loss of geometry to be permanent movements of the track foundation. The track foundation comprises of 3 layers; the ballast around and immediately beneath the sleepers; the deep ballast and/or underlying construction layers and the subsoil. The bottom two layers provided they have been adequately specified and dimensioned will only give a significant direct contribution to the permanent movement during the early life of the track or when there has been an increase in the loading.

2. The top ballast is that which is subjected to the highest stresses and which is being constantly disturbed by track maintenance and traffic and it is this layer which is the subject of greatest interest as it is applicable to all track in all locations.

3. The present investigation is into the loss of the vertical longitudinal track geometry and is in two main parts; one part examining the factors influencing the deterioration of the track is being done through the development of a computer model which simulates the deterioration of the track caused by various factors. The other part is concerned with the parameters used in the model which influence the deterioration and in particular those parameters which influence the movement of the ballast.

4. By the successful development of a model which can

take into account all variations in traffic and track parameters it is possible to use this work to predict the effect on track deterioration of changes in the traffic and also of new proposals for the design of the track. From the predicted track deterioration the resultant changes in the cost of track maintenance can be evaluated, enabling the economic viability of proposals to be determined at an early stage. The other value of this work is the achievement of a greater understanding of the mechanisms causing the deterioration of track geometry and the role the various track parameters play in this deterioration. The design and specification of the track and of the methods used for maintenance and renewal can then be improved and optimised for a given traffic or for economic constraints.

THE DETERIORATION OF TRACK GEOMETRY

5. Before developing a model for the track deterioration and determining the most relevant parameters which need to be considered, it is important to obtain an understanding of the behaviour of the track geometry with time. Over the past few years both on British Rail and on other railways considerable information has been collected on the deterioration of the vertical profile.

6. Routine maintenance on most railways is carried out by combined tamping and lining machines. After rectification of the geometry by these machines the loading from the traffic causes settlement in the ballast and as this settlement is not uniform faults in the track geometry develop. It has been found that the geometry deteriorates very rapidly at first and then more slowly. The standard deviation of the track profile from an idealised filtered profile has been chosen to describe the quality of the geometry and this is normally given for a 200 m section of track. Using these values the changes in track geometry during a maintenance cycle can be clearly seen <u>Fig 1</u>. After the improvement given by the tamping machine

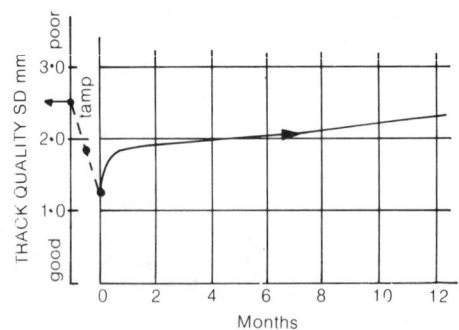

FIG. 1 CHANGE IN VERTICAL TRACK
QUALITY DURING A TYPICAL MAINTENANCE CYCLE

there is an extremely rapid period of deterioration which is normally completed within two months. After this there is a slow rate of deterioration which is approximately linear with time or traffic. On average the improvement given by the machine and the subsequent deterioration under traffic is found to be a function of the quality of the track. Fig. 2

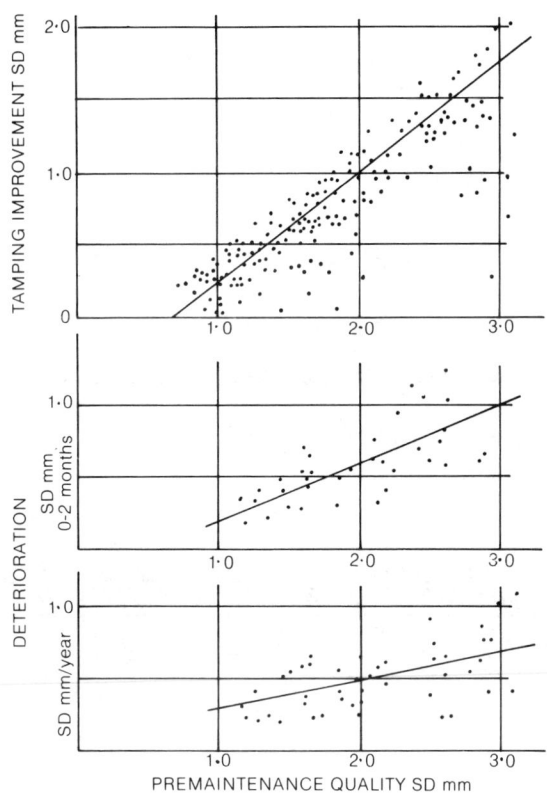

FIG. 2 THE IMPROVEMENT IN TRACK QUALITY PRODUCED IMMEDIATELY BY THE TAMPING MACHINE AND THE DETERIORATION CAUSED BY TRAFFIC (FOR THE FIRST 2 MONTHS AND THE SUBSEQUENT RATE FROM 2 MONTHS TO A YEAR)

7. Another important observation on the deterioration of the vertical geometry is that a given section of track always has a tendency to deteriorate towards a given shape Fig 3. Observations of many kilometers of track has led to the conclusion that in general good track remains good and poor track remains poor throughout a period of many maintenance cycles, and the number of tamping operations has very little influence on this quality Fig 4. In this figure a diamond represents the track quality for a 1/8 mile section of track. The height of the diamonds show the range

of values recorded over a 5 year period. On the two sections taken as examples of good and poor track the arrows indicate each tamping operation. The conclusion from this work is that the track has an "inherent quality" which was determined during the early part of its life and that this is a function of the quality of the components of the track and its foundation and of the quality of the work on

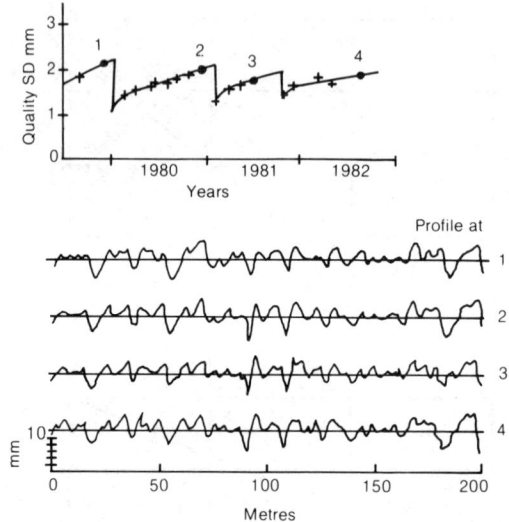

FIG. 3 THE VERTICAL PROFILES OF A SECTION OF TRACK TOWARDS THE END OF FOUR MAINTENANCE CYCLES

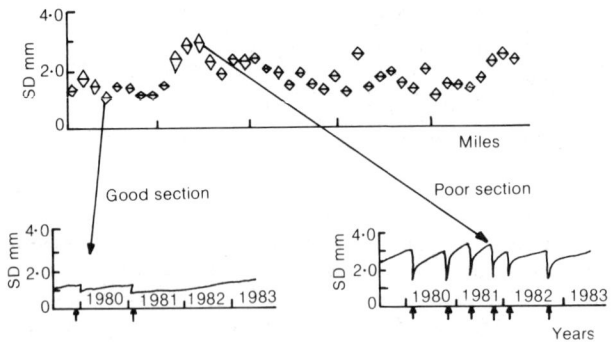

FIG. 4 VARIATION OF VERTICAL QUALITY OVER SEVERAL YEARS

installation.

THE CAUSES OF DETERIORATION

8. The deterioration of the vertical geometry can only be caused by the differential settlement of the ballast and the foundation. However differential settlement can result from different sleepers experiencing different loads or from differential settlement of sleepers under identical loads. Examination of these causes leads to the conclusion that there can only be six mechanisms of deterioration; these can all take place simultaneously and are often interactive. The six mechanisms are:

1) Dynamic Forces

   A vertical load from a wheel can be continously varying along the railhead due to either irregularities in the running surface or irregularities in the vertical geometry. This will cause sleepers at different locations to experience different loads and even if the ballast bed is uniform those sleepers which consistently experience the highest loads will settle the most causing track faults.

2) Rail Shape

   Defects in the longitudinal shape of the rail (a lack of straightness) which can arise from the manufacturing process or from misalignments produced at the welds. Even with a force from an axle which remains perfectly constant and with a uniform ballast bed the lack of rail straightness results in some sleepers carrying different loads from others The result is a tendency to gradually impress the 'free' unstressed shape of the rail into the ballast bed with the passage of successive axles. After an infinite number of axles the loaded track profile will be the same as the original free rail shape. The rate at which these faults develop is a function of the wavelength of the fault; those with a wavelength of less than 5 metres appear rapidly whilst those having a wavelength greater than 15 metres would not normally appear within a tamping cycle. Detailed investigations into this phenomena have been carried out (Refs. 1 & 2) comparing track measurements and theoretical predictions.

3) Sleeper Spacing

   If within a section of track the sleepers have a variable spacing then those which are closer together will individually carry less load and therefore settle less under the traffic and conversely those which are further apart individually each carry a greater load and settle more. However within the limits of spacing which occur on a modern main line the contribution from variable sleeper spacing is unlikely to be of much significance.

4) <u>Sleeper Support</u>
If the elastic support seen by the individual sleepers is different then the load carried by individual sleepers is different. A sleeper having a stiff support will carry more load than the adjacent sleepers on softer supports and this will cause that sleeper to have a tendency to settle more.

5) <u>Ballast Settlement</u>
As ballast is a random arrangement of stones, under a given load the settlement is also random. It is obvious that if all other conditions are uniform then there will still be deterioration of the track geometry due to the differential settlement properties of the ballast. As well as the intrinsic lack of homogeneity of the ballast there are other factors which would contribute to differential movement such as the action of the tamping machine and the variability of the sub-grade; these will be dealt with later in the paper.

6) <u>Substructure</u>
The final contribution to the differential settlement of the foundation and the deterioration of the geometry is that which comes from the deeper ballast, sub-structure layers and from the sub-grade. Normally most tracks have an adequately designed foundation and thus movements arising directly from settlement in these layers are negligible in the short term when considering the standard maintenance cycle.

9. Whilst the discussion above has examined the six causes in isolation it is obvious that these are interactions to be considered. The most important of these is that any irregularity occurring in the vertical geometry from the last 5 causes will, if the traffic is travelling at a reasonable speed, result in a variation in dynamic load. This will increase the effect due to the first cause.
British Rail during the course of the investigations into the deterioration of the track geometry have developed computer programs which simulate the deterioration of a section of track (Ref. 3). The dynamic loads on the section of track, are calculated and from the maximum load seen by each sleeper the settlement of that sleeper can be calculated. The resulting deteriorated shape of the track can then be used to re-calculate new dynamic loads and by repetitive cycling of this procedure the development of track deterioration can be simulated.

10. This programme has been used to evaluate the 6 modes of deterioration in isolation (i.e. neglecting any interactions) and the predictions can be seen in <u>Fig. 5</u>. Each of these examples shows the development of the deteriorated shape up to 500 000 passes of a 20 tonne axle and in each case the variations in the parameters used are those which could be typically encountered.

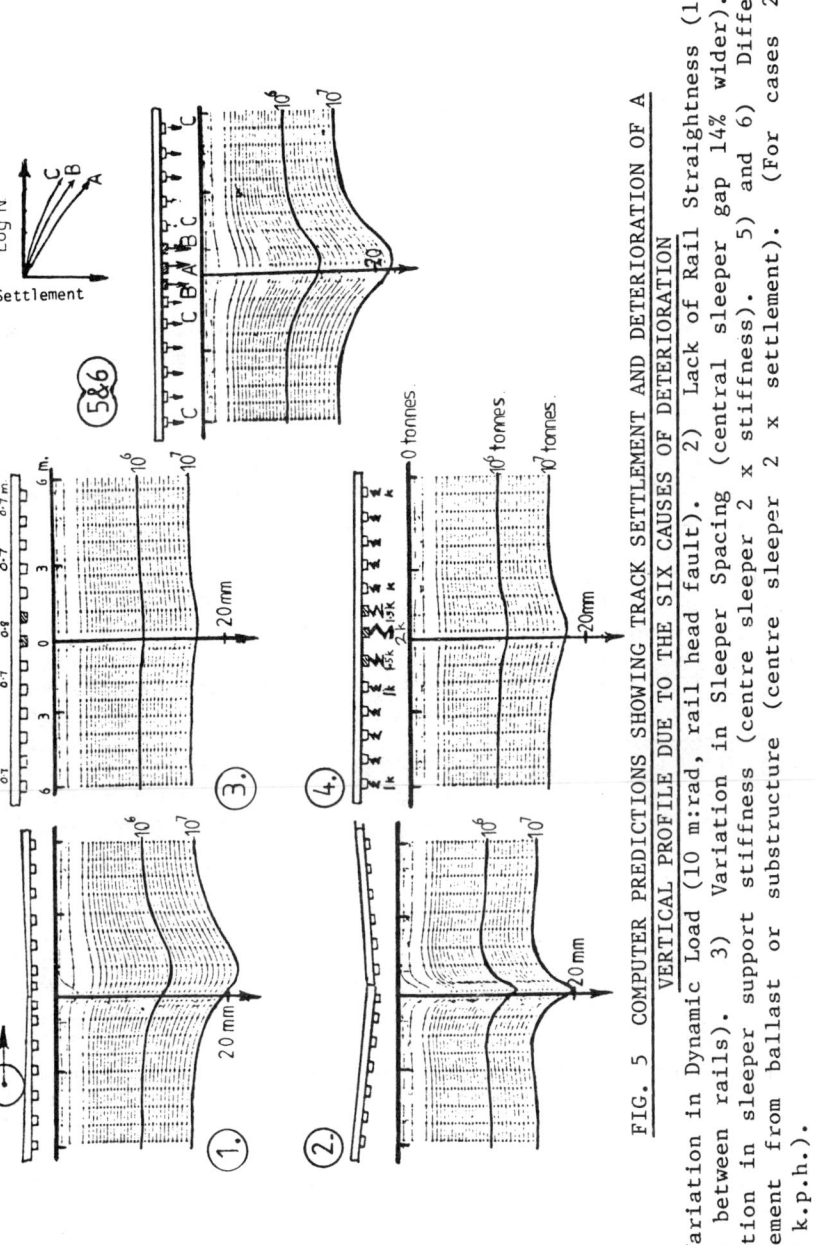

FIG. 5 COMPUTER PREDICTIONS SHOWING TRACK SETTLEMENT AND DETERIORATION OF A
VERTICAL PROFILE DUE TO THE SIX CAUSES OF DETERIORATION

1) Variation in Dynamic Load (10 m:rad, rail head fault). 2) Lack of Rail Straightness (10 m:rad angle between rails). 3) Variation in Sleeper Spacing (central sleeper gap 14% wider). 4) Variation in sleeper support stiffness (centre sleeper 2 x stiffness). 5) and 6) Differential Settlement from ballast or substructure (centre sleeper 2 x settlement). (For cases 2 to 6 v = 0 k.p.h.).

11. It can be seen from these examples that the main contribution to track deterioration comes from the dynamic loads, the rail shape and the variability of ballast settlement; the contribution from the variability in sleeper spacing and the direct effect of support elasticity appear to make only a minor contribution.

12. Further work is continuing in the development of these models and in the collection of data relevant to their use with the aim of producing a universal model which will be able to predict deterioration from a representative combination of causes.

13. There are two points worthy of note which arise from the study of deterioration. The first is that in none of the track observations or theoretical calculations has a "runaway" condition been observed where the rate of deterioration (per unit of traffic) increased with time; the second is that all the theoretical considerations suggest that the track will always tend towards a shape which is stable when the loads have been re-distributed among the sleepers such that the rate of settlement of all the sleepers is uniform. This condition is normally achieved after an extremely large number of cycles so although it is a true measure of the "intrinsic" quality of the track it rarely exists in practice.

BALLAST SETTLEMENT

14. The previous section identified the three main causes of the deterioration of vertical geometry as the dynamic loads, the rail shape and the performance of the ballast. A detailed examination of the first two are outside the scope of this paper so the second part of this paper considers those factors which influence the settlement of the ballast and thus have direct bearing on the deterioration of the vertical profile. The first thing to consider is the general form of the law governing the settlement of the ballast. It is a well known fact that after tamping the ballast settles rapidly and then the rate of settlement decreases with increasing traffic. It is often assumed that the settlement of the track is proportional to the logarithm of the number of axles or total tonnage, however examination of all available data from world wide sources (Refs. 4, 5, 6) indicates that whilst this may be considered as a reasonable approximation over a short period of time a significant underestimation can occur for large numbers of axles, Fig. 6. Each of the lines on this figure shows the average settlement for a site where measurements were made and covers a wide range of traffic and track conditions. Laboratory and field experiments have shown that a better approximation is obtained when the settlement is considered to be proportional to the fifth root of the number of axles. This agrees well with the site measurements up to 10^6 load cycles but does tend to underestimate the settlement above this value. The best fit is obtained when

an equation of the form

$$S = K_1 N^{0.2} + K_2 N$$

is taken, where $K_1 N^{0.2}$ predominates up to 1 million load cycles and K_2 is a very small factor such that $K_2 N$ only becomes relatively significant above this value.

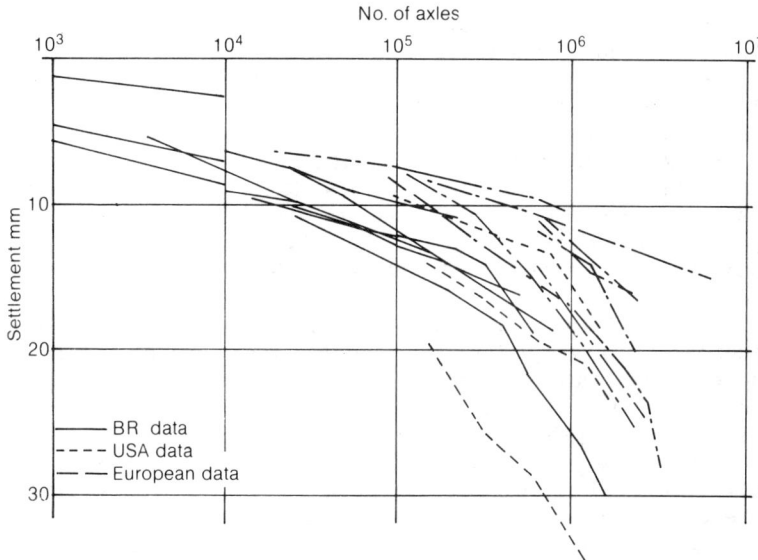

FIG. 6 AVERAGE SETTLEMENT OF TEST SITES

The hypothesis is put forward that the first factor represents the movement of the disturbed ballast following tamping whilst the second factor represents the residual settlement which is occurring in the deeper ballast and foundation and which although probably also obeying a law of the form $S = N^{0.2}$ from the day the track was relayed, can be considered as linear over the length of a later maintenance cycle.

15. In addition to the number of load cycles there are many other factors which influence the total amount of settlement.

A) The load on individual sleepers which is controlled by
    1. Axleload
    2. Rail Section
    3. Sleeper Spacing
    4. Track & Foundation Elasticity

B) The traffic loads
    5. The dynamic loads
    6. Mixed Axleloads

C) The settlement of individual sleepers which is influenced by:-
    7. The size and type of sleepers

8. Ballast type and condition
9. Sub-ballast and sub-grade
10. Type of track maintenance
11. Lift given during maintenance

Axleload is probably the most predominant factor. It has been found that for slow speed tests under a rolling load that the settlement is directly proportional to the axleload. Earlier tests in the laboratory with tri-axial samples of ballast (Ref. 7) and later tests with model sleepers also confirm that a linear relationship between axleload and settlement is a reasonable approximation. No direct confirmation of this has been found from track measurements as these are always complicated by the mixture of axleloads which occurs on running lines.

16. The earlier tri-axial tests (Ref. 7) showed that it was the higher loads which predominated the track settlement and that loads below 50% of the maximum commonly occurring load had no influence on settlement. This was even valid when the number of low loads were a significant proportion (90%) of the total, thus for many lines it is only the locomotive axles which are causing the track settlement. The earlier tests with mixed axleloads have been re-examined and have led to the conclusion that the best approximation which can be made is to assume a law where the rate of settlement depends on the axleload and the total settlement. This law can be used to derive an "equivalent axleload" A_e where:-

$$A_e = \left(\frac{A_1{}^5 N_1 + A_2{}^5 N_2 + A_3{}^5 N_3 + \text{------}}{N_1 + N_2 + N_3 + \text{-------}} \right)^{0.2}$$

for a mixture with N_1 axles having a load of A_1 etc. A summary of this equation to give the equivalent axle load for a simplified mixture of 15, 20 and 25 t axleloads is shown in Fig. 7. An investigation is in hand to check the validity of this concept and although this is showing that the law tends to exagerate the difference between two traffics it is the best approximation which is currently available.

17. The influence of rail section and sleeper spacing was also examined with the rolling load rig and these showed that their influence could be taken into account simply by calculating the load carried by

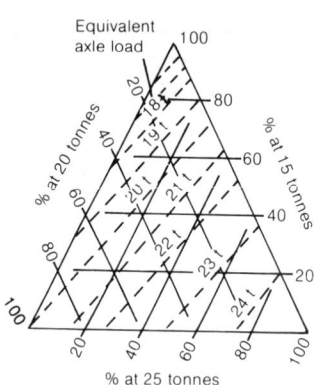

Equivalent axle load

% at 20 tonnes

% at 15 tonnes

% at 25 tonnes

FIG. 7 ESTIMATION OF EQUIVALENT AXLE LOAD FROM A MIXTURE OF THREE DIFFERENT AXLE LOADS

each sleeper. **Fig. 8** shows the influence of different sleeper spacings and rail sections on the average load carried by a sleeper. The change in settlement can thus be estimated by multiplying by the load factor. The direct influence of dynamic loads was investigated using tri-axial samples (Ref. 7) and these tests showed that for a given level of load the frequency of loading did not influence ballast settlement (up to the test limit of 40 Hz).

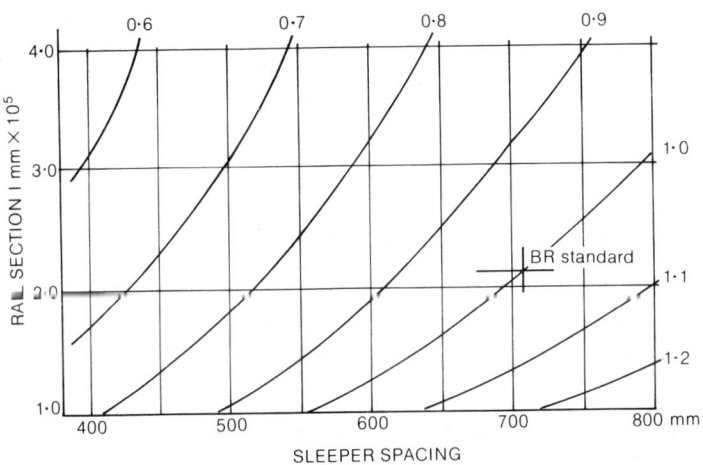

FIG. 8   INFLUENCE OF SLEEPER SPACING AND RAIL SECTION
ON THE LOAD AND SETTLEMENT OF A SLEEPER

For a nominal load applied to a sleeper one of the most important factors is the lift given by the tamping machine, the larger the lift given to the sleeper the greater will be the settlement under the subsequent traffic. **Fig 9** shows this relationship for a number of test sites and indicates

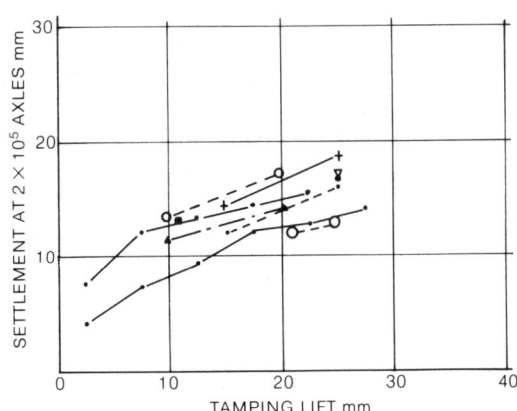

FIG. 9   INFLUENCE OF LIFT GIVEN BY THE TAMPING
MACHINE ON AVERAGE TRACK SETTLEMENT

that the average settlement after a 20 mm lift is approximately 1.1/2 times the settlement after a 5 mm lift.

18. The influence of the type and size of sleeper has also been the subject of investigation, for type, the available evidence is not very consistent but there are indications that the settlement of timber sleepers may be between 70 and 90% of that of the equivalent sized concrete sleepers. Extensive tests on half scale sleepers to determine the influence of shape and size has shown that the settlement is inversely proportional to the support area with a slight influence of shape factor (length: breadth) i.e. a square support area is the optimum. This finding confirms the tentative results drawn from track measurements (Ref. 6).

19. Other factors influencing settlement which have not yet been fully quantified are that of ballast type; where tests have shown that the settlement of a very poor limestone can be twice that of a good granite; and the influence of the sub-grade or sub-ballast. Recent measurements have shown an increase in track settlement of 1.5 times with a change in track stiffness from 90 to 70 kN/mm and laboratory tests also indicate the lower the ballast support stiffness the greater the settlement.

20. The results of all these investigations can be used to derive a general equation which quantifies the settlement of ballast with a reasonable degree of accuracy:-

$$S = K_s \frac{A_e}{20} \left((0.69 + 0.028L)N^{0.2} + 2.7 \times 10^{-6}N \right)$$

where Ae is the equivalent axleload, N the total number of axles L the lift given by the tamping machine and K_s is a factor which is a function of sleeper type and size, ballast type, and of the sub-grade; for a normal B.R. track $K_s \triangleq 1.1$

CONCLUSIONS

21. The results of this work to date have shown that it is possible to have extremely good track with low rates of deterioration i.e. having a good "inherent" quality. To achieve this it is necessary to have straight rails and ballast with uniform settlement properties. Control of these two factors will naturally control the variations in dynamic loads for a given traffic. In order to improve the "inherent" quality of new and existing tracks work is now being undertaken to produce straighter rails and also to straighten rails in-situ. The work on the settlement of ballast has identified the attention which must be paid to having a consistent track foundation with a good and properly prepared ballast bed which has sufficient depth to limit the elastic deflections and variability of the sub-grade. On existing tracks that part of the inherent quality caused by differential ballast settlement can be significantly reduced by selective packing under the sleepers; for example by the Pneumatic Ballast injection machine now under development. For new construction and

relaying,improved methods of preparing a consistent ballast bed are under investigation.

22. Research into the deterioration of the track and the behaviour of ballast has resulted in these developments which will result in future tracks with a better quality and requiring less maintenance.. In addition when completed the work will enable such factors as the size of sleepers, rails etc. to be optimised to give the most economic solution for the first and maintenance costs of the track and to evaluate changes in vehicle design.

ACKNOWLEDGEMENTS

The author would like to acknowledge his many colleagues in the B.R. Research Department at Derby for their valuable contributions in conducting tests and developing computer programmes and also the Office de Recherches et d'Essais (O.R.E.) de l'U.I.C. - Utrecht, whose D117 work programme provided additional data. He would also like to thank the British Railways Board for permission to publish this paper.

REFERENCES

1. ROUND, D.J. The effect of rail straightness on track geometry and deterioration - Rail Technology, Proceedings of a Seminar - Nottingham University 1981.

2. FREDERICK, C.O. The effect of rail straightness on Track Maintenance. Conference on Advanced Techniques in Permanent Way Design - Madrid 1981.

3. LANE, G.S. The effects of track and traffic parameters on the development of track vertical roughness - Proceedings of the Conference of heavy haul Railways - Colorado 1982.

4. BOSSERMAN Tie Ballast Interaction, Proceedings of FAST Conference, FRA 1982.

5. SELIG - E Theory of Track Maintenance Life Prediction. D.O.T. Report 1981.

6. ORE D117 RP2 Optimum adaptation of conventional track to future traffic - Utrecht 1973.

7. SHENTON, M.J. Deformation of railway ballast under repeated loading conditions - Railroad Track Mechanics and Technology - Pergamon Press 1978.

Discussion on Papers 17–21

PROFESSOR DR -ING G. BERNSTEIN, University of Karlsruhe

The future of rail passenger transport in the next decade lies
in high speed. Therefore analysis and research must be
concentrated on train speeds of 200-250 km/h. It is
impossible for civil engineers to wait for mechanical and
machine construction engineers to build rail vehicles with
reduced vibration.

Mr Griffin and Mr Stanworth, what does your analysis
indicate about the vibrations from passenger and freight
trains at speeds greater than 100 km/h?

What are your proposals for the construction of a
permanent way on existing lines with mixed traffic?

MR GRIFFIN and MR STANWORTH

You are correct when you say that trains are likely to go
faster, although it is unlikely that goods trains, which seem
to cause the most noticeable vibrations, will be required to
travel at the highest speeds which you suggest.

If we suppose, however, that the speeds of existing
trains over existing track were to be doubled, the
consequences for the vibration dose

$$(\int_o^T a^4(t) \ dt)^{1/4}$$

can be seen, at least qualitatively. Force levels on the
ground would increase and wayside vibration would be likely to
increase similarly. The vibration dose is dominated by the
peak acceleration levels; if these doubled (while the time of
passage halved), we would expect the dose to change by a
factor of approximately $(16/2)^{1/4} \approx 1.7$. This, however,
neglects the fact that the acceleration spectrum would have
changed, so that both the weighting function and the dynamic
behaviour of the observation point would have had some
influence on the dose. If (subject to the same limitations)
the vibration dose were to remain the same, the acceleration
levels would have had to be allowed to increase by a factor of

only $2^{1/4} \approx 1.2$, suggesting that dynamic force levels on the ground could only have been allowed to increase by the same amount.

Since the train speed would have been doubled, the longest track feature 'wavelengths' of significance would have increased to 7 m, although sleeper pitch would have lost some of its significance.

It therefore seems that, on the basis of ground vibration perception, only very modest increases in dynamic ground force can be entertained as train speed rises, and that the significant roughness wavelengths increase in direct proportion to the speed.

For the classical mixed traffic railway we suggest that vehicle design will have to play a substantial part in the limitation of ground vibration as speed is increased.

MR C. F. BONNETT, Director of Civil Engineering, London Transport

The recommendations contained in Paper 18 for the use of a floating slab support for underground rapid transit tracks are interesting but raise a number of fundamental issues.
Firstly, in many instances, notably on the London underground, the existing tunnels are very tight and floating slabs could not be accommodated in the tunnel diameters. For new work tunnels would have to be larger and as the cost increases with at least the square of the diameter the overall increase in cost could be considerable. Secondly, to increase the tunnel diameters on existing tunnels would involve high costs which could not be justified.

Within these 'facts of life' for an existing railway, Professor Eisenmann, what do you recommend for reducing noise and vibration?

Incidentally, the extension to the Piccadilly line to serve terminal four at Heathrow will have elastic boots fitted to the sleepers for part of its length. These boots have been accommodated within an internal tunnel diameter of 3.70 m.

PROFESSOR J. EISENMANN

On the London underground with a ballastless track (overall height of about 30 cm) it is difficult to reduce noise and vibration. All that can be done is to decrease the elasticity of the rail support and rail grinding. If the deflection of the rail under the running wheel is 1.5-2 mm the damping is about the same as for a ballasted track.

MR B. I. SINGAL, Design Manager, Mass Transit Railway Corporation, Hong Kong

The Hong Kong Mass Transit Railway adopts special floating

track bed at the more sensitive locations. The delegates at this conference may be interested in our findings.

The measurement of vibration levels at the tunnel inverts, at the level of the platform and at ground level above the tunnels as well as measurements of the noise level in buildings adjacent to the tunnels have shown that the intrusive ground-borne vibrations occur mainly in the frequency range 63–125 Hz. This is true both for tunnels founded on soft ground and in rock. The velocity level of the vibrations is below 0.1 mm/s.

The floating track bed is provided when the background noise level is exceeded by more than 10 dB. The mass–spring system as shown on the simplified Fig. 1 has been adopted, so far, at four locations. The weight of the floating track bed is about 2 t/m and the vertical natural frequency of vibration lies between 11 and 15 Hz with a vertical deflection of 3–5 mm under loaded conditions. The overall height between the rail level and the underside of the floating track slab is about 600 mm. Both continuous and discontinuous floating track slabs have been used depending on programme and access constraints.

The designs are providing satisfactory attenuation of vibrations.

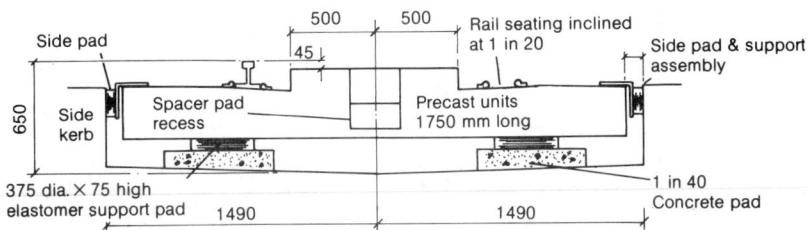

Fig. 1. Floating track bed in Hong Kong

MR R. M. RENFREW, Executive Director, Canadian Institute of Guided Ground Transport, Kingston

The recommended solutions (dual-block, rubber-isolated track bases) are very expensive to install. They have been used in several North American underground systems with variable results. They are not generally effective at the more troublesome seismic frequencies (greater than 20 Hz) because the energy levels are very high and the residential buildings have resonances at these frequencies.

The chief reason for considering spring–mass systems is the deficiencies of the typical bogie. Many European bogies are monomotor type with high stiffness primary suspensions. These bogies have poor curving characteristics which excite the rail by a high level of work input. Owing to the stiff

primary suspensions there is some coupling of the intermediate
truck and motor masses into the axle mass, also resulting in a
high vertical energy input. The suspensions also produce
corrugations in the rails in relatively short times, further
aggravating the noise and vibration generated.

The PCC streetcar is an outstanding example of an
effective, low vibration suspension design. Why have European
designs neglected the system level solution? The success of
the best high speed train designs, which operate at high
dynamic load conditions, illustrates the potential for good
bogie design to reduce substantially the need for expensive
track designs for urban transit systems.

MR F. I. MAU, Civil Design Engineer, BHP Engineering, Sydney

Care must be taken in the use of track moduli for design.
Professor Raymond has discussed the importance of this factor,
while others differ. In Fig. 2 of the paper it seems that the
effect of the modulus diminishes away from the underside of
the sleeper.

In addition, the numerical values of the moduli given
differ from experience in Australia. Therefore, when putting
moduli into the design, consider carefully the validity of the
values. I suggest for example that experience in Australia
where track moduli on ballasted track do not exceed 50 MPa for
the best quality concrete-sleepered track should be
considered.

PROFESSOR G. P. RAYMOND

You point out that the moduli for concrete-sleeper-type track
that I used for illustration differ from those experienced in
Australia. You then suggest that calculations for a track
modulus of 50 MPa be presented. In response it should be
noted that the solution used for the concrete sleeper
calculations with stiff pads relates to an actual case history
and the effect of changing to softer pads is clearly
illustrated and shows quite a dramatic drop in track modulus.
Extensive testing of sleeper pads has been performed by Dean
et al. [1] and some typical results are shown in Fig. 1. If
the ballast, sub-ballast and/or subgrade are altered to a
material with lower pseudo modulus the track modulus would
also be lowered. This is demonstrated in Fig. 2 using test
data by Bosserman [2] from FAST where the only variable is
stated to be the ballast type and size. Australian practice
is to use soft pads with concrete sleepers so it is
understandable that the higher values in my paper are not
applicable. They may be regarded as a warning to those
contemplating the use of hard pads on stiff ballast, sub-
ballast and subgrade. As to the calculation of examples using
a track modulus of 50 MPa my paper clearly outlines the

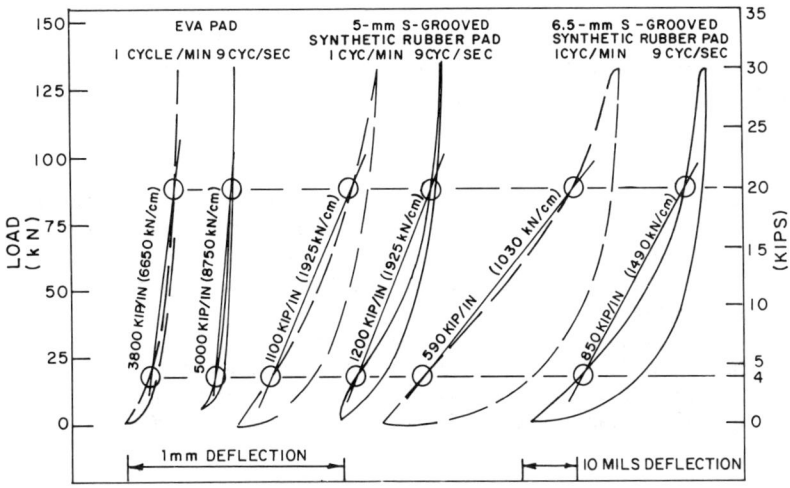

Fig. 1. Typical load-deflection of sleeper pads

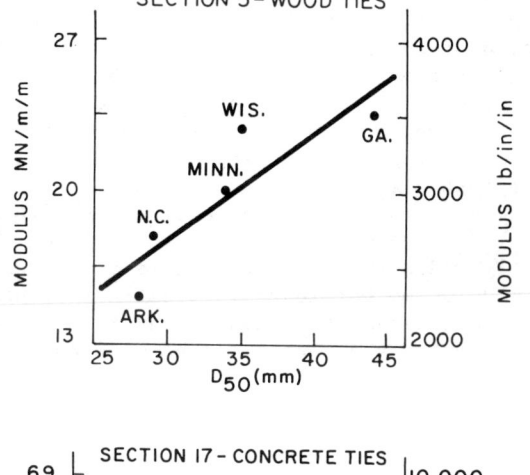

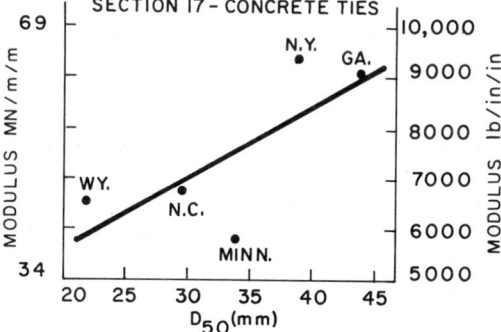

Fig. 2. Effect of variation in ballast particle size on measured track modulus

procedures related to any track modulus the reader wishes to analyse. In this regard all the calculations were performed on a portable microcomputer costing less than 2000 Canadian dollars.

REFERENCES

1. Dean, F. E., Ahlbeck, D. R., Harrison, H. D. and Tuten, J. M. Effect of tie pad stiffness on the impact loading of concrete ties. Proc. 2nd Int. Heavy Haul Railway Conf., Colorado Springs, September 1982, paper 41, pp. 442–458.
2. Bosserman, B. N. Ballast experiments at FAST. Proc. FAST Engng Conf., 1981. US Department of Transport Report No. FRA/TTC-82/01, pp. 45–54 (available from National Technical Information Service, Springfield, Virginia).

MR D. J. AYRES, Soil Mechanics Engineer, British Railways Board, London

I have encountered various approaches to track bed design while collaborating with my counterparts on other railways in the production of the new UIC Bulletin No. 719R. These approaches are either based on some strength or modulus measurement or on soil classification methods. I favour the latter empirical approach but note that all methods take account of the traffic loading spectrum. In paragraph 17 Professor Raymond states that the loading is greatest beneath the rail; however, in my experience of cases of bearing capacity failure, the maximum depression of the subgrade is never directly under the rail but occurs to the cess (field) side about 100–150 mm horizontally from that point. It is associated with a heave about 1.4 m away from the running edge and this heave acts as a counterweight when cumulative deformation occurs. Experiments to load this heave have prevented further loss of level of rail, although this is not a practical remedial measure. It indicates that stress comparison with soil strength is not the sole criterion for rail movement as the mechanism of failure is also relevant.

Movement of the subgrade under traffic loading is not continuous but intermittent in response to rainfall. It is an effective stress failure with subgrade strength depending on the position of the water-table and on the capillary suction in the subgrade soil. Water reaches the subgrade and reduces the soil strength by satisfying the negative pressures before it reaches the side drain. However, the subgrade can tolerate a small amount of rainfall and, if an impermeable layer could deflect water to the side, stability ensues. This has been British Rail practice since 1957, when polyethylene film was placed as a sandwich in the filter layer (blanket) placed over cohesive subgrades. Over 1000 miles of this have been installed with satisfactory results. The implication is that good drainage is required for the ballast bed under and around the sleepers but that the sub-ballast layer should be impermeable.

The known static filter criteria are not adequate for dynamic conditions over cohesive subgrades and produced the British Rail specification for granulometry on the basis of sands known to have filtered successfully in track for up to a century. Filter sand over clay showed slurry penetration of up to 5 mm and I feel that a minimum thickness of sand of 10–20 mm is necessary to dissipate the local dynamic pressures in clay slurries immediately above the interface. In practice much thicker layers of sand are needed but these can be kept to a minimum of, say, 150 mm or even, with care, to 100 mm if a geotextile is placed between sand and ballast as a separator. A programme of laboratory tests carried out by British Rail for ORE has shown that silty clay slurry under dynamic loading will pass through all the commercially marketed geotextiles. The experience of the ORE subcommittee on drainage is that a geotextile alone will not provide a permanent filter.

For good separation over sands, I favour a non-woven, fully heat-bonded geotextile of minimum 350 g/m^2; although this would puncture it would not disintegrate at depths beyond the effect of the tamper.

PROFESSOR G. P. RAYMOND

In general, your comments complement the points that I have made. My comments will thus be confined to minor differences of opinion. You question the statement that maximum stresses are below the rail seat since heave occurs (during subgrade failures) 1.4 m away from the track. Failures, however, occur after the small depression caused by these maximum stresses have allowed water to accumulate and generally to soften the underlying subgrade. Once subgrade softening occurs failure results at the weakest combination of soil strength and applied stress. This could be at any location not necessarily, nor generally, the point of maximum stress.

My experiences with membranes below the track are that they must permit, or be installed with a granular soil layer which permits, horizontal (lateral) drainage. I fully agree that it is unacceptable to use a granular soil layer that is sufficiently coarse grained that it permits penetration of silt and/or clay size fines. Such material clearly cannot act as a filter or separator to the subgrade. In this regard the use of fine-sized sand combined with a geotextile is mentioned in my paper.

DR -ING J. EISENMANN, Professor of Civil Engineering,
Technische Universität München

We have done a series of laboratory tests to investigate
ballast behaviour under a repeated load at 3 Hz up to
1 000 000 cycles superimposed with 50 Hz vibration. Another
test series has been conducted with 50 Hz loading using a
large vibrator (vibrogir). The test results for the German
Railways show that during the first 10 000 cycles (about
200 000 tons in the track) there is a typical break-in phase
with heavy settlement. After that there is a consolidation
that is linear on a logarithmic scale. We found much scatter
in several tests with the same parameters. An increase in the
load leads to an increase in the settlement. The same effect
takes place when the ballast is watered. The test results are
in good agreement with the well-known track deformation. The
laboratory tests will be used as input for a computer program
for modelling track behaviour.

I agree with Professor Raymond, Dr Shenton and the
speakers of Session 1 with regard to the ballast behaviour
during the break-in phase. This emphasizes the necessity of
reconsidering the tamping technology. A step in this
direction could be the Stoneblower.

BIBLIOGRAPHY
Eisenmann, J. and Kaess, G. Das Verhalten des Schotters unter
Belastung. Eisenbahntechnische Rundschau, (1979), No. 3.
Eisenmann, J. Verhaltensfunktion des Schotters.
Eisenbahningenieur, (1981), No. 3.

DR S. L. GRASSIE, Research and Development Engineer, Pandrol,
London

The papers that have been presented have addressed the
function of ballast as a resilient layer. Ballast also plays
a significant role as the principal source of damping in the
track. At frequencies of excitation above about 150 Hz, the
resilience of ballast in track with concrete sleepers and rail
pads usually becomes insignificant compared with the
resilience of the rail pad. This is because there is a
significant increase in the relative movement of the rail and
the sleeper on the rail pad. However, the damping which
ballast provides is critical: the premature cracking of
sleepers in sections of track with poor ballast arises
primarily because the vibration of the sleepers is almost
undamped.

We have found that for track in its usual condition there
is relatively little difference in the stiffness and damping
characteristics of different ballast. There is evidence that
tamping reduces both, albeit temporarily. However, in a
recent experiment, conducted coincidentally during cold
weather, we found that the frozen ballast was much more highly

damped than ballast in its usual state. This finding is consistent with a hypothesis that damping arises from wave transmission through the ballast, which would be facilitated if the ballast were frozen. Although it is impractical to freeze all our ballast, investigations should be made of those characteristics of ballast which enhance its damping performance and could thereby increase the life of concrete sleepers under dynamic loading.

In paragraphs 12-14 of Professor Raymond's paper, reference is made to the 'dynamic wheel load' arising from a wheel flat. It should be stressed that the force calculated is only the low frequency component of the dynamic wheel load, referred to as the P_2 force by Frederick and Round (Paper 12). The high frequency component of the impact force, the so-called P_1 force, is usually a more severe component of the force between the wheel and the rail. However, because this force is reacted primarily by the rail and the sleeper, its effects on the ballast are negligible.

The P_1 force usually causes more damage than the P_2 force to the rail and the sleepers. It can be reduced by reducing the effective mass of the system. This can be done most effectively by decoupling the rail from the sleeper using a resilient rail pad.

BIBLIOGRAPHY
Grassie S. L. and Cox S. J. The dynamics of railway track on unsupported sleepers. Submitted to Proc. Instn Mech. Engrs, Ser. D.

MR S. T. LAMSON, Research Associate, Canadian Institute of Guided Ground Transport, Kingston

One of our current projects is to simulate with computer models the increase in track roughness with traffic accumulation. We have found that the general logarithmic relationship between permanent ballast settlement and axle load cycle greatly underestimated the change in track roughness. Therefore I am pleased to see the same observation from the latest British Rail research results, Dr Shenton, and would be interested to have your explanation for it.

DR Y. SATO, Railway Technical Research Institute, Tokyo

With regard to ballast settlement (Paper 21, beginning at paragraph 14), two points appear in the paper: settlement of the ballast itself and settlement of the ballast owing to that of the subgrade. Although settlement of only the ballast itself occurs for a solid subgrade, the settlement of the subgrade becomes the dominant factor for a poor subgrade.

Through experiment and experience in Japan, the settlement of the ballast itself has been characterized as follows.

(a) The settlement y is expressed as

$$y = \gamma [1 - \exp(-\alpha x) + \beta x]$$

where γ is the magnitude of initial settlement, α is the duration of the initial settlement, β is the rate of growth of the settlement and x is the number of loading cycles. Fig. 1 is a graphical expression of this formula.

(b) Here, α is related to the initial conditions, such as tamping, and β is related to the long-term settlement. Then, β is expressed as $\beta \propto$ (tie pressure) x (ballast acceleration).

(c) The tie pressure is proportional to the axial forces, the ballast acceleration, vehicle characteristics and the train speed.

(d) Thus the ballast settlement is proportional to the passing tonnage, the train speed and the characteristics of the vehicle.

Repeated number of loads

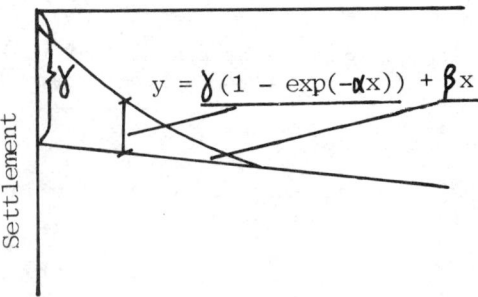

$$y = \gamma(1 - \exp(-\alpha x)) + \beta x$$

Fig. 1. Settlement of ballast

However, the settlement of subgrade is of the fatigue type as shown in Fig. 6 of Paper 21. In this case, it is important to specify the kind of soil, the ballast pressure and the water content.

DR M. J. SHENTON

In reply to the comments of Dr Eisenmann, Mr Lamson and Dr Sato concerning the laws relating to ballast settlement, the equation is definitely not semilogarithmic and shows no discontinuity at 10^5 cycles. Both the equation given in the paper and that of Dr Sato are continuous functions tending towards linearity with the number of axles for very high

numbers of axles. However, there are differences between
these two equations for the earlier part of the curves.

One of the difficulties in interpreting settlement data
is that there are very few tests in which sufficient readings
have been taken during the early part of the test (below 1000
cycles) and which extend in excess of 10^6 cycles. More recent
laboratory tests have met this criterion and have confirmed
that a law of the form $S = K_1 N^{0.2}$ covers the full range of
load cycles. During these tests it was only the ballast layer
which was deforming permanently.

It is interesting to note that British Rail have found in
both laboratory and field tests that a law of the form
$D = KN^{0.2}$ also relates to the horizontal displacement D of a
sleeper to the number of applied horizontal loads N. This
confirms that although earlier tests, normally in the
controlled stress environment of the triaxial test, indicated
a logarithmic law the behaviour of ballast 'in situ' is
different, and the logarithmic law would give an underestimate
when extrapolated to large numbers of cycles.

MR W. I. JONES, Chief Civil Engineer, New Zealand Railways,
Wellington

The main conclusions of Paper 21 include the importance of
geometrically straight rails. Rails cascaded from a heavy
traffic line to a lighter line are usually of lower geometric
accuracy than new rails. If the rails are straightened, Dr
Shenton, is there a problem with reduced fatigue life, arising
from the plastic deformations needed to straighten the rails?
Is it common to roller-straighten cascaded rail before re-
laying?

DR M. J. SHENTON

It is not normal to roller straighten cascaded rail on British
rail.

Assuming no change in the applied loads, rail
straightening could reduce the fatigue life of some thermit
welds depending on the position of defects. However, in most
cases (if properly done) rail straightening, in conjunction
with grinding when required, will reduce the dynamic loads
thus increasing the life. Work is currently under way on this
topic at the research department.

PROFESSOR K. RIESSBERGER, Technical University of Graz

During the conference the shortcomings of the present track
behaviour have been repeatedly pointed out, but it is worth
noting that the tamping machines and tamping methods are not
responsible for all of them.

The tamping machine is recognized throughout the world as the backbone of mechanized track maintenance, resulting in major savings in costs and manpower. It is used effectively on all high speed lines and heavy haul tracks. The main problem of track stability is not the small rapid settlement illustrated in some of the diagrams, but the conditions of ballast and drainage. On German Railways, where these conditions have been brought up to a good standard, the necessary tamping cycles have been reduced to only once in four years on the heavily used lines and as much as once in seven years on others.

The stone blowing machine is still only in its development stage and none of the promising claims made by several of the speakers have yet been proven in day-to-day practice. To make the promises into facts is a strong challenge.

As Dr Shenton stated, care in the initial track laying to ensure a high inherent quality is of the utmost importance to minimize the subsequent maintenance requirements. Also, ballast cleaning is considered a type of track laying. The use of modern ballast cleaners fitted with devices to control the depth and the slope of the cutter bar enable an almost perfect formation level to be obtained, which is a fundamental requirement of good track and it is desirable to fit such devices to all existing ballast cleaners. Together with it is the requirement that reasonable lengths of track be cleaned in one operation.

Surprisingly, nothing has been said about the insertion of gravel layers between the soil and the ballast. It should be noted that for this complicated work also mechanized methods are at hand. Therefore this technically well-proven solution can be carried out at a reasonable cost.

As stated by Mr Frederick, the straightness of rails and welds is equally important for good track. Problems of this type can be solved in situ with the STRAIT procedure. This is provided by a specially developed tamping machine which firstly bends the rail (or weld) plastically, next tamps the four sleepers on each side and then grinds the spot to remove the uneven uplifted surface. Thus all sources of impact are removed, which otherwise would hammer the weld into the dipped position again.

Professor Selig has referred to the dynamic track stabilizer. This machine was built to improve the settlement behaviour of newly tamped track. The concept of compacting by high initial loads to minimize further settlement by the subsequent traffic has also been illustrated by Professor Raymond. The aim has not been fully achieved, because of weight limitations, but further refinement should enable the machine to fulfil the initial expectations.

Points about the stabilizer not mentioned include the facts that

(a) lateral resistance is increased such that full speed can be achieved immediately after track maintenance, even on the 270 km/h Paris-Lyon line

(b) vertical stiffness is equalized

(c) longitudinal stresses are smoothed out effectively and thus the hazard of track buckling is much reduced

(d) uniform track gauge is automatically achieved because vibration shakes the rails to the outer limits of the fastening.

Several of these machines are in satisfactory operation.

In general the track technology in the next decade will be dominated by mechanized methods and the already existing principles will prevail.